ROVER P5 AND P5B

Other Titles in the Crowood AutoClassics Series

ROVER
P5 and P5B

The Complete Story

James Taylor

First published in 1997 by
The Crowood Press Ltd
Ramsbury, Marlborough
Wiltshire SN8 2HR

British Library Cataloguing-in-Publication Data
A catalogue record for this book is available from the British
Library.

ISBN 1 86126 003 2

Picture Credits
Mark Dixon, *Classic & Sportscar* magazine, Dave Bull, British
Motor Industry Heritage Trust, Dan Clayton, Author's collection,
Tony Poole, Tony Cope, Sigvard Heringa, Michael Mayr-Harting,
Mme H. Chapron and the National Motor Museum.

Typeface used: New Century Schoolbook.

Edited and designed by
D & N Publishing
Membury Business Park, Lambourn Woodlands
Hungerford, Berkshire.

Printed and bound by The Bath Press.

Contents

Acknowledgements

I am pleased to record my thanks to the many former employees of the old Rover Company who have willingly shared with me over the last 20 years their recollections of direct involvement with the P5 and P5B models. They are listed here with their former job titles or job descriptions. Sadly, some of them are no longer with us, and their names are marked by an asterisk (*).

David Bache*	Chief Stylist
Graham Bannock	Market Researcher
Gordon Bashford*	Chief Engineer, Advanced Vehicles
Joe Brown	Design Engineer, Advanced Vehicles
Lou Chaffey	Service Mechanic in the Competitions Department
Dan Clayton	P5 Project Engineer
Jim Dodsworth	P5 Coupé Project Engineer
William Martin-Hurst*	Managing Director
Bruce McWilliams	President, Rover Motor Company of North America
Harry Mills	Photographic Department
Tony Poole	Styling Department
Frank Shaw	Chief Engineer, Transmissions
Ken Stansbury	P5 and P5B Project Engineer
Jack Swaine	Chief Engine Designer
Brian Terry	Test Engineer, Experimental Department Engine Section
Dave Wall	Engine Section Design Engineer
Philip Wilson	Experimental Department

In addition, particular help came from Jon Backhouse (Rover Service Newsletters), Stan Banting (former Rover Company historian), Richard Brotherton (British Motor Industry Heritage Trust), Henk Bruers (2.6-litre P5s), Dave Bull (P5B literature), Madame H. Chapron (Chapron convertible), Anders Clausager (British Motor Industry Heritage Trust), Tony Cope (South African P5s), Mark Dixon (colour photography), Ian Elliott (editor of *Freewheel,* the Rover Sports Register's magazine), H.S. Fry (former Director of FLM Panelcraft), Spen King (former head of New Vehicle Projects at Rover), George Hansson (Panelcraft convertible), Sigvard Heringa (Graber convertible), Nick Mathias (Rover P5 Owners' Club), Michael Mayr-Harting (2.4-litre P5 models), Harold Radford (Radford divisions) and Karam Ram (British Motor Industry Heritage Trust). Last but not least, innumerable P5 enthusiasts have contributed little pieces of information to the jigsaw over the years. They know who they are.

Introduction

I have been interested in the Rover P5 and P5B models for more than twenty years now, and writing this book has given me a wonderful opportunity to make sense of all the information I have collected in that period. The P5s led on to my interest in other Rover products and in the four-wheel drives from Solihull, but they remain my favourites.

I still remember my first test-drive in a 1963 Mk II Saloon, and how – despite the body filler and the garish lower-half respray – I was completely captivated by the car's spaciousness, refinement, and olde-worlde wood and leather charm. From that car, I graduated to an early 3.5-litre Coupé which I cherished for nine years until some low life removed it from the station car park where I had trustingly left it. There has almost always been at least one 3-litre or V8-engined car in the Taylor fleet ever since, and at the time of writing the tally appears to be four. It has been a most rewarding addiction, and one of which I hope I shall never tire.

Nevertheless, I think there are valid reasons besides my own self-indulgence why the P5s and P5Bs deserve a book recording their story. The most obvious one perhaps is that the cars now have a substantial following in the classic car movement, at least in Britain. But from a longer-term historical perspective, we should recognize that the P5B was the last of a dying breed of car. The traditional British luxury saloon probably saw its greatest period in the fifties, but the Rover picked up this great tradition and carried it right through to the early seventies. By then, other practitioners of the art had fallen by the wayside, and it was the Jaguar concept of a luxury saloon as an altogether more sporting car which had found acceptance worldwide. Rover could never have directly replaced the P5B because the car's market had dwindled to almost nothing – but that did not stop many people from feeling a twinge of regret for its passing.

In the nineties, there may be no room for a car with the characteristics of the P5s and P5Bs, but today's Rover Group has openly acknowledged its heritage by borrowing styling cues and features from these big saloons for its latest products. The wood-and-leather interiors of today's Rovers are a deliberate evocation of the calmer days of the P5s, their radiator grilles are quite openly copied from the distinctive style worn by the P5s, and even the chromed surround of the Rover 600's rear number-plate was inspired by the styling of that part of the P5. The cars may be gone, but they have certainly not been forgotten.

James Taylor

Chronology

1958   P5 introduced as Rover 3-litre

1961   Revised Mk IA 3-litre models introduced

1962   Mk II 3-litre models and new Coupé variant announced

1963   Rover 2000 (P6) introduced

1964   Rover 95 and 110 (P4) models ceased production

1965   Mk III 3-litre Saloon and Coupé introduced
         Rover acquired manufacturing rights to Buick V8 engine

1967   3.5-litre Saloon and Coupé (P5B) models replaced 3-litres

1968   Model names changed to 3½-litre Saloon and Coupé
         V8 engine introduced in 3500 (P6B) saloons

1970   Range Rover introduced with 3.5-litre V8 engine

1973   Production of 3½-litre Rovers ended

1   Design and Development

Strange though it may seem, a great deal about Rover's P5 luxury saloon can be attributed to the success of the company's utilitarian multi-purpose Land Rover. Without that, Rover would not have been in a position to consider a second saloon range alongside the existing P4 models in the early fifties, and because of it, the P5 which was announced in 1958 was very different indeed from the car which Rover's directors had envisaged in 1953.

The Land Rover's success had actually taken Rover by surprise. Conceived in 1947 and put into production in 1948, the vehicle had always been intended as a temporary product to keep Rover in the motor manufacturing business until the market for its traditional gentleman's cars – disrupted by

the Hitler War – returned to normal. However, things did not quite work out the way Rover had planned. The continuing need to export ensured that the saloon market would never again be the one Rover had known in the thirties, and the extraordinary success of the Land Rover meant that the whole focus of Rover's business activity changed. Throughout the fifties, Land Rovers sold in twice the quantities of the Rover saloons, and the Rover Company was constantly struggling to keep up with demand. Once the Land Rover had become established, Rover actually became a maker of light utilities which had a profitable sideline in saloon cars – although its management was always very reluctant to view the company in such a light!

In the early fifties, the Rover Board of Directors pursued two related policies. One was to reinvest profits from the Land Rover in new products, and the other was to find a volume product which might replace the Land Rover when the Land Rover bubble burst. In the first few years of the new decade, the new products came thick and fast: the original P4 75 saloon was joined in 1953 by 60 and 90 derivatives, while the original 80in wheelbase Land Rover was replaced at the same time by two new models, offering alternative wheelbases of 86in and 107in. Rover also looked at ways of building additional models around existing major components, and among these were the convertible P4 projects of the first half of

---

Why P5?

The question of Rover's post-war model designations has never been resolved to everyone's satisfaction, but it seems pretty clear that the 'P' stood for 'Post-war.' The first new product envisaged in 1945 (but never made) was a miniature car known as the M-type and sometimes called the M1. That seems to have taken care of the first number, and the second number went to the P2, which was a warmed-over pre-war car. The 1948 P3 and 1949 P4 followed logically after that, and the P5 became the fifth model in the sequence.

the decade and the plan for a half-car, half-Land Rover hybrid called the Road-Rover.

By 1953, however, engineering chief Maurice Wilks had started to think about a new Rover saloon, smaller and cheaper than the P4 and sold in much higher volumes. He probably reasoned that this would meet all eventualities: if Land Rover sales held up, then Rover would be able to afford to build it as an additional model, and if they collapsed, the new car might actually replace the Land Rover as the company's high-volume product. He thought of this car as the P5, its code-number being the natural choice to follow on from the existing P4 saloons, and he expected it to be not just one car but a whole range which would eventually replace all the P4 models.

It was round about this time that David Bache joined the company as its first stylist – although no such grand title yet existed at Rover, and Bache was simply the member of the Drawing Office with particular responsibility for the exterior appearance of Rover products. Many years later, Bache recalled that Maurice Wilks then envisaged the car as a 2-litre V6 of 14ft overall length, a sort of Rover version of the Lancia Aurelia. Drawing up proposals for its exterior appearance was Bache's first major job after joining Rover, and he rose to the occasion, preparing two scale models. Many people at Rover who saw them were very impressed indeed. Eventually, Maurice Wilks himself came to look at them, walked round them carefully, congratulated Bache on his excellent work, and then dropped the bombshell: 'These are wonderful pieces of work, Mr Bache, but I'm afraid we can't produce them.' 'Why's that, sir?' asked Bache, crestfallen. 'Because Rovers should be discreet, and these two are head-turners,' answered Wilks.

The 'small' P5, 1952–1954

Maurice Wilks' original idea was that P5 should be a smaller and cheaper car than the existing P4, about the same size as the contemporary Vauxhall Velox or Ford Zephyr Six and aimed at roughly the same kind of customer. By 1953, Wilks and his deputy, Robert Boyle, expected P5 to replace the P4, probably after an initial period in which the two were available alongside one another.

Wilks' initial specification for what was always called P5 was drawn up late in 1952. Its details varied over the next two years or so, but the size of the car remained fairly constant, with a wheelbase set initially at 103in and later increased to 104.5in. The P5 was always intended to have Olaf Poppe's base-unit construction, the base-unit itself being manufactured at Solihull and the skin panels being supplied by Pressed Steel. There were to be rear-wheel drive and independent rear suspension, and weight was to be kept down to between 22cwt and 24cwt.

During 1953, the plan was initially to use a 2.6-litre V6 engine, but later on weight was reduced and there was to be a 2-litre V6 with the option of a 1.8-litre version of the existing four-cylinder engine; Wilks and Boyle even intended to allow room for the forthcoming Land Rover four-cylinder diesel engine! By May 1954, the engines were conceived as a 2.5-litre V6 and a twin-carburettor edition of the existing 2-litre four-cylinder. However, it was over the summer of that year that the original P5 concept was abandoned and the car became a larger machine with more powerful engines and a monocoque bodyshell instead of the base-unit.

No running prototypes of the 'small' P5 were ever built, although some versions of the V6 engine did run in P4 development cars between 1952 and 1954, and a full-size wooden mock-up of the car was made in the late spring of 1954.

An important influence on P5 styling was the special fixed-head coupé built on a P4 90 chassis by Pininfarina in Italy. It was actually the second of two cars built, the first being a drophead coupé with the same styling. The car originally had right-hand drive but was converted to left-hand drive when it was sold to Rover's Spanish importer. It was pictured in Spain by the National Motor Museum's Nick Georgano. The original drophead coupé and a later copy of it by Mulliner's of Birmingham still survive, but the fixed-head car has disappeared.

Bache was of course not the only one who put forward proposals for the new P5. On the engine side of the Drawing Office, chief designer Jack Swaine was asked to dust off designs for a V6 engine he had been looking at in the late thirties. The Research Department obtained a Lancia to examine. And the head of Production Engineering, Olaf Poppe, suggested that the new car should dispense with the traditional separate-chassis construction and should have a rigid steel-skeleton framework to which running gear and unstressed skin panels should be bolted. This was revolutionary thinking at the time – probably too revolutionary for a company as staid as Rover – but it predated by a couple of years Citroën's use of the concept in the DS19 of 1955. Rover, which carried the idea through

into the P6 Project which began in 1956, was then accused of plagiarizing the Citroën idea when the P6 entered production in 1963!

Nevertheless, this 'small' P5 was doomed not to progress beyond the drawing-board and mock-up stage. Once again, the Land Rover was the cause. Production was running at such high levels that it was filling every available corner of Rover's Solihull factory, and component production was filling the company's numerous satellite factories in the Birmingham area. The question of whether a high-volume saloon car might go into production does not even seem to have reached the Board of Directors for a decision: it must have been quite obvious that Rover simply did not have the room to build such a car, regardless of the project's other attractions.

11

The influence of the Farina car's styling is obvious on this full-size mock-up of the 'small' P5. The picture dates from September 1954, by which time the mock-up would have been redundant because Rover management now wanted P5 to be an altogether bigger car. Note the heavy visor, which was a David Bache idea.

The second-stage P5 was a much larger car than originally planned, and one of its engines would have been a new 3-litre V6. This full-size mock-up was pictured on 8 May 1955, and shows that the basic styling of the production P5 had already been established. There is still a peaked visor over the windscreen, however, and the bodyside trim strip still echoes Pininfarina's work.

CHANGES OF DIRECTION

Some time during the summer of 1954, the original 'small' P5 Project came to an end. However, the fact that Rover could not contemplate building a high-volume car did not alter the fact that P4 would one day need to be replaced, and for the next couple of years, P5 seems to have been envisaged as a replacement range for P4. Meanwhile, Rover was going ahead with plans to expand the P4 range upwards and so it was only natural that the P5 range should have embraced the same spread of variants as P4. At its bottom end was a four-cylinder 2-litre car – equivalent to the Rover 60 – while at its top end was a 3-litre V6 which would have positioned the car just above the twin-carburettor 105S version of the P4 which was then under development.

This second stage of the P5 project lasted for about two years, from the summer of 1954 until some time in 1956. A September 1954 photograph of the first-stage styling mock-up was captioned in the Rover archives as showing a 'proposed 1957 P5' – in other words a car intended for introduction in autumn 1956 as a 1957 model – but those who worked at Lode Lane in the mid-fifties can remember no strict timetable and no sense of urgency to get the car into production. And in fact, the ideas for P5 seem to have continued to evolve during this period, with the result that a third – and this time, definitive – stage followed.

Things had been allowed to drift for too long. During 1956, there was a major reorganization at Rover, when Robert Boyle took over from Maurice Wilks as the company's Engineering Chief and Wilks himself moved up to become joint Managing Director alongside his brother Spencer. Peter Wilks, a nephew of both the elder Wilks, was appointed as Assistant Chief Engineer under Boyle, and under him came four Project Engineers – the one in charge of P5 being Chris Goode. At the same time, David Bache was taken out of the Drawing Office and was put in

---

The Wilks brothers

Between the thirties and the start of the British Leyland era, Rover was dominated by two figures — the brothers Spencer and Maurice Wilks. Towards the end of that period, their nephews Spen King and Peter Wilks occupied key positions in the company, too. William Martin-Hurst, who became Managing Director in the sixties, was also related to the Wilks family by marriage. People who worked at Rover during that period have nothing but kind words to say of the friendly and productive atmosphere engendered by this family of remarkable engineers.

Spencer Wilks, the elder of the two brothers, was Joint Managing Director at the Hillman company when that company was taken over by the Rootes brothers in 1928. He joined Rover in 1929 as its General Manager, and in 1933 was promoted to Managing Director. It was he who was largely responsible for creating the image of the Rover as a discreet and well-made car for the professional man. In 1957, he took over as Chairman, but retired from that post in 1962. He nevertheless retained his seat on the Rover Board, and became the company's life President in 1967. Spencer Wilks died in 1971.

Maurice Wilks had also worked at Hillman, and he followed his brother to Rover, becoming the company's Chief Engineer in 1931. He remained Rover's technical chief — although the title of the job changed — until becoming its joint Managing Director in 1956. Even then he remained very influential on Rover engineering, and continued to do so after taking over from his brother as Chairman in 1962. Sadly, he died just a year later.

charge of his own staff in a new section called the Styling Department.

Robert Boyle and Maurice Wilks (who never really relinquished his hold on Rover engineering) also took a long hard look at the various projects under way in the Engineering Department at this time. The Road-Rover, another project which had been drifting aimlessly since the early fifties, took on a radical new direction. The need for an eventual P4 replacement was re-addressed in a new project known initially as the PX Project, which began in September 1956 and was later renamed the P6 Project. That left the P5 Project, which was redefined. From now on, P5 would be a low-volume luxury model, positioned above the P4 in the Rover range.

A number of important decisions followed. The first was that P5 would not have a range of engines, as originally envisaged, but rather a single engine which would be larger and more powerful than those in the P4s. That almost automatically meant that it needed to be a 3-litre engine – and it was probably at this point that the car became known by its eventual public name of the Rover 3-litre. And yet Maurice Wilks was unwilling to abandon the idea of a whole range of P5s. If the car was to have only one engine, it could nevertheless have two different bodies. Just as the Rovers of the thirties and forties had been available with upright six-light saloon bodies and more rakish four-light Sports Saloon bodies, so Maurice Wilks decided that the P5 should have both staid saloon and rakish Sportsman bodywork. Early in 1957, work started on this new Sportsman model which Wilks initially hoped would be released

By September 1955, when this picture was taken on the sports field at Rover's Solihull factory, the full-size mock-up had been modified at the front and had a less severe-looking visor. The quarter-lights in the front and rear doors had gone, and David Bache was experimenting with slim window frames made of bright metal. The banked-up grass conceals the wooden model's underframe.

alongside the standard saloon. In fact, it did not enter production until 1962, by which time it was known as the Coupé model.

DEVELOPMENT IN DETAIL

Throughout this period, Chris Goode remained P5 Project Engineer, the man responsible for coordinating the work of the various departments which were involved with designing and developing the P5. His task cannot have been made any easier by the fluid nature of the whole project, and by the constant changes in the car's size. Goode had also been Project Engineer on the original 'small' P5 Project, and he had seen its wheelbase size go from 103in to 104.5in, from there to 106.5in in autumn 1954, up again to 108.5in in November 1954, and finally to the 110.5in agreed for production! Even then, as David Bache remembered, Maurice Wilks asked for the car to be made two inches wider at the last minute.

With the end of the 'small' P5 Project in 1954, the Rover engineers also changed their ideas on the car's structure. Gordon Bashford, who was then in charge of design research, had for some time been investigating Olaf Poppe's base-unit idea, but there was still a lot of work to be done before he could recommend it for production. Besides, part of the original scheme had been that Rover should build that base-unit at Solihull, and it was now clear that there would be no space for that because of continuing high demand for the Land Rover. The P5's body would therefore have to be built by an outside supplier, and the most obvious choice was Pressed Steel, who were already building the P4 bodyshells.

Pressed Steel were then highly-regarded in the British motor industry, and their expertise with monocoque body structures was second to none. However, they had no experience at all of base-unit structures. Their advice to Rover – most probably in the person of Gordon Bashford – would certainly have been to use a monocoque structure for the P5. They had recently worked with Jaguar on the monocoque for that company's compact 2.4-litre saloon, and had recommended that

The P5's monocoque was built by Pressed Steel, initially at that company's Swindon body plant. This is actually a 1961 example, but differs only in minor details from the earliest shells.

the engine and front suspension should be mounted on a separate subframe, bolted to the bodyshell, so that the front of the shell would not need heavy reinforcement. The Jaguar, announced in 1955, had been a huge success, so it was no surprise that Pressed Steel suggested Rover should go down the same route with the P5.

Rover had no experience of monocoque structures at this stage, and quickly bought for evaluation an example of the car generally acknowledged to be the best big monocoque

saloon in Europe – the Mercedes-Benz 220S. No doubt a Jaguar 2.4-litre was also examined at Solihull. Gordon Bashford, meanwhile, took on the job of developing a detachable front subframe for the new Rover. In true cautious Rover fashion, it ended up being heavily over-engineered, with massive box-section side-members and cross-members which were all of three inches deep.

Bashford also took responsibility for the new car's suspension design. Independent rear suspension faded from the picture when

A very early P5 monocoque displays the complexity of its front end structure. Much of the rigidity in the assembled car came from the subframe which bolted to this and carried the engine, gearbox and suspension.

the second-stage P5 Project began in autumn 1954, probably on the grounds of complication and cost, and instead the plan was to use a variation of the P4's progressive-rate semi-elliptic leaf springs on a beam axle. For the front axle, however, Bashford decided not to have the heavy coil-spring-and-wishbone suspension of the P4 but rather to have a lighter suspension using laminated torsion bars as the springing medium. He had been looking at this type of front suspension as early as 1952, when it

was one possibility being entertained for the original 'small' P5, and had now become convinced of its merits. Nevertheless, his assistant Joe Brown later recalled that they had a lot of trouble getting the torsion bar suspension to work properly.

Brakes and steering more or less chose themselves. The servo-assisted all-drum brakes developed for the 1956 model P4 90 were extremely good by the standards of the time, and there was no reason not to use them again for the P5. Disc brakes certainly

This is the rear end of a very early P5 monocoque, showing the original style of bumper mounting brackets. The boot floor was flat and wide, but there was not quite as much space as there appears to be in this picture because the fuel tank went in over the rear axle.

Top of the Rover range after September 1956 were the automatic 105R (pictured) and the 105S (with overdrive gearbox) variants of the P4 range. They had twin-carburettor engines of 2.6 litres which gave 108bhp and around 95mph. These were well-respected cars in their day, but by comparison, David Bache's P5 proposals looked incredibly sleek and modern.

were under investigation at Rover – and they were used on two of the experimental gas turbine cars in the fifties – but they were not yet ready for production. Nor were they really expected: the first British production cars with disc brakes did not appear until 1957, and both of those (the Triumph TR3 and the Jaguar 3.4-litre) were sporting machines quite unlike the new Rover. The existing Burman recirculating-ball steering on the P4 models was also perfectly adequate at a time when luxury-car customers did not expect power assistance, and so it was retained for the P5.

Transmissions were more complicated, however. The earliest schemes for the second-stage P5 car envisaged no manual gearbox at all, but rather a range of automatic or semi-automatic transmissions. Several such designs were then under development by Frank Shaw's transmission engineers, and one of them – the Roverdrive – went into production during 1956 on the P4 105R. Dan Clayton, who joined the P5 project team as a Technical Assistant in 1957, remembers that a Roverdrive transmission was actually

used in the first prototype or prototypes of the P5, but that a decision was soon made to buy-in the Borg Warner DG automatic instead. This American-designed gearbox was being made in Britain and had already proved its worth in Jaguars, among other marques. It also had the advantage of sapping rather less engine power than the Roverdrive notoriously did. By 1956, if not earlier, a decision had also been taken to offer the P5 with a manual transmission. This would simply be the existing Rover four-speed gearbox, fitted with the Laycock de Normanville overdrive which had become optional on the 1956 model P4 90 and was an expected feature on luxury cars such as the P5 was now intended to be.

For the second (1954–1956) stage of the P5 Project, Jack Swaine developed a 3-litre version of his 60-degree V6 engine. This was to be the engine in the most expensive versions of the P5, and its design target was 120bhp at 5,000rpm. Prototypes were built and did run, but the engine team had considerable difficulty with them. The main problem was breathing, and Swaine says now that he and his colleagues were forced to accept that the oversquare, short-stroke architecture of the 2,950cc V6 did not marry up well with the inlet-over-exhaust valve configuration which had been designed in the thirties for a long-stroke engine. The engine simply could not be persuaded to rev high enough, and the best Swaine ever saw was 100bhp at a low 4,000rpm. This was just not enough and so development was abandoned, probably some time in 1956.

Rover still needed a 3-litre engine, however. So Swaine asked one of his staff in the Engine Drawing Office, Norman Bryden, to develop a 3-litre engine from the 2.6-litre straight-six which was then the staple of the P4 range. The existing engine had already been bored-out once, and that had entailed moving the bore centres further apart in the block so that there would be a water jacket all round. This redesigned engine had entered production as recently as 1953, but its block did not allow room for the even larger bores which Bryden needed for the 3-litre engine. So the block was redesigned yet again, with the bores spaced even further apart.

At the same time, Bryden took the opportunity to change the four-bearing crankshaft for a seven-bearing design, which he knew would improve smooth running and expected also to improve engine durability. This feature gave the engine its internal code-name of 3L7 – 3-litre, seven-bearing. He also incorporated the latest roller-type cam followers, originally designed for the Land Rover diesel engine and more durable than the pad-type followers in the existing Rover sixes. The resulting engine gave 115bhp at 4,500rpm – not enough to give the heavy P5 startling performance but enough to make it respectably quick. In due course, a short-stroke 2.6-litre version of the engine was developed for the P4s, and it replaced the earlier 2.6-litre type in 1959 – just a year after the P5 entered production. This had the advantage of simplifying machining operations at Rover and so saving manufacturing costs, as all the six-cylinder engines then had a common design of block and head.

While all this engineering work was going ahead, David Bache was sketching up ideas for the P5's exterior appearance. The overall shape he drew up was not unlike some 1955 model Chryslers, and other American cars no doubt had an influence. Nevertheless, Bache was also paying close attention to what continental European manufacturers were doing, and among his influences must have been the 'Ponton' Mercedes, the Lancia Aurelia GT, and the Alfa Romeo

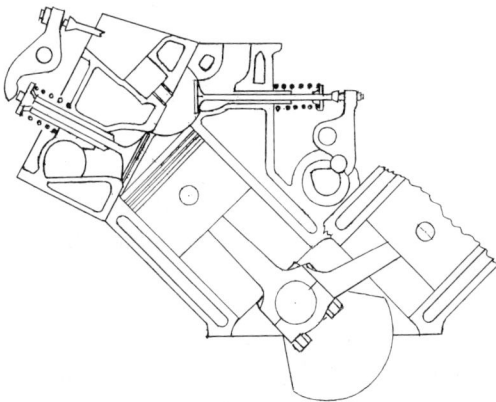

The largest engine in the second-stage P5 was to be a 3-litre V6. No pictures of the engine have been located, but designer Jack Swaine remembered it well enough to produce this sketch some 40 years later for Ian Elliott, editor of the Rover Sports Register's magazine, Freewheel. *The bore was 3½in (88.9mm) and the stroke 3 ⅛in (79.37mm) to give 2,950cc. The cylinder banks were angled at 90 degrees to one another and the firing intervals were at 90 degrees and 150 degrees. There were two SU carburettors, one at the front of each cylinder bank.*

Jack Swaine and the stillborn V6 engines

Jack Swaine had become Rover's Chief Engine Designer by the time the P5 was on the drawing-board, but his influence on the P5 can be traced back much further. Swaine joined Rover in the thirties, taking over from Robert Boyle (who left for a time to work at Morris Engines and then returned to Rover to become Maurice Wilks' deputy and later took over from him as Chief Engineer).

Some time in 1934 or 1935, Maurice Wilks asked Swaine to look at a V6 engine; his idea was that such an engine could be fitted into the same engine bay as an in-line four-cylinder, and thus Rover would not have to produce two different chassis in order to offer both four-cylinder and six-cylinder cars. As the existing Rovers had narrow bonnets, the V6 could not have a wide angle between its cylinder banks, and so Swaine drew up an engine with a 60-degree vee. To keep the width down still further, he tipped the angle of the cylinder head towards the centre of the engine, and for simplicity, he used a single camshaft in the centre of the vee, acting directly on side exhaust valves and through pushrods to overhead inlet valves.

Although the V6 did not go into production, this sloping-head, inlet-over-exhaust valve design gave such good results that Swaine used it again for the post-war in-line engines which he designed in 1946–1947. The 3-litre straight six which went into production in 1958 was a direct descendant of these engines, and used the same characteristic valve configuration and sloping cylinder head.

When the P5 project began in the early fifties, Swaine dusted off his pre-war V6 designs, and a 60-degree V6 figured in early plans for the car.

1900, all of which had been announced in 1953. In addition, the fixed-head coupé on a P4 chassis which Maurice Wilks had commissioned from Farina in Italy a couple of years earlier left its mark on the P5, most obviously on the wide grille. At a later stage of the project, as Bache happily acknowledged many years afterwards, he borrowed the idea of the P5's wraparound windscreen and dog's-leg front door cut-outs from the 1954 Facel Vega.

Ultimately, however, whatever Bache proposed had to get past the eagle-eye of Maurice Wilks. As Bache himself later admitted, he occasionally had to bite his tongue in frustration when Wilks toned down some of his ideas, and even though Bache had become Rover's Chief Styling Engineer in 1956, the final shape of P5 still owed a lot to Maurice Wilks' ideas on how a Rover should look.

Tony Poole, who left the Gas Turbine Department to join Bache in the new Styling Department in 1956, remembers witnessing the styling development of the P5, initially from afar and then at first-hand. During 1954 and 1955, there had been various scale models. 'They were very Americanized,' remembers Poole. 'Nearly all had got visors, peaks around the front, loads of chrome.' The first full-size mock-up he saw had been built in a small area divided off from the Experimental Shop by tin sheeting, which was the only place available to the stylists before they were given their own department in 1956. The full-size mock-up was made of wood, as the American practice of using clay had not yet reached Rover. Poole also remembers the transition from the 'small' P5 of 1952–1954 to the larger car which followed: 'I came to work one morning, a Monday, and there was this notice stuck to the mock-up. The Wilks brothers had been in over the weekend. And the notice simply said, DO NOTHING ON THIS MODEL 'TIL WE ARRIVE. That was the stage when they decided to go big.

We put in two inches in the wheelbase and about an inch and a half in the width.'

THE FIRST PROTOTYPES

The basic outline of the P5 nevertheless came together quite quickly, and two hand-built prototypes were made towards the end of 1956. Decorative details had still not been settled, of course, and the cars had a mixture of makeshift items (such as side and tail lights) and production P4 components (such as bumpers and, probably, door handles). Brian Terry, who was then a test engineer working under Jack Swaine on engines, remembers the cars as 'two matt-grey, very uninspired P5 prototypes, which were very light and went like the blue blazes compared to the production one! One was a manual and one was an automatic.'

All the development work was carried out on these two cars, and no more experimental prototypes were built. Problems which showed up included exhaust boom, and Dan Clayton remembers that a number of experiments were conducted with different systems before a satisfactory solution was found. The back seat ride was also poor, and Clayton remembers that the problem was traced to the rear spring mountings, which were swinging shackles like those used on P4s and Land Rovers. Chris Goode's solution was to design a fixed mounting with some inbuilt compliance – the distinctive triangular mounting plate with its Metalastik bushes which eventually entered production.

The two experimental prototypes initially ran on trade plates, and some footage of them testing at MIRA was used in the promotional film about the 3-litre which Pathé made for Rover in 1958, *In the Rover Tradition*. The second car – numbered EXP/3L/2 – was road-registered as VWD 129 on 27 March 1957 probably so that it could go on mileage testing. One car was used for static rig tests and the other, remembered Brian Terry, was parked in the open under a window of the Engine Test Department until it simply fell apart from rust. The Rover archives reveal that EXP/3L/2 – which could have been this car – was scrapped in December 1966.

A delightfully informal shot of stylist David Bache at his desk, round about the time the P5 was being designed.

By about April 1957, the P5 styling mock-up had taken on its definitive dimensions, even though decorative detail had still not been settled. The dog's-leg front door cutout had now arrived to allow a bigger windscreen wraparound, and David Bache knew where he wanted the side trim strip to go, even though the front end detail was not yet clear. Rover never had any intention of calling the car a '3L7' – that was the engine designation, used simply to demonstrate what wing badging might look like. Indicators have not yet been settled, and appear to be in those lamps on top of the wings, no doubt inspired by the parking lamps fitted to contemporary Mercedes saloons. The small model in the background is a Road-Rover, with a three-way colour split; Tony Poole remembers the two lower colours as being pink and black! The Styling Department had its own viewing studio complete with turntable by this time.

The fate of the other is not recorded, but it would certainly have been broken up.

The next P5s to be made were six 'production prototypes' – cars built using production components – which were assembled over the summer of 1958. Most of the problems which Rover encountered on these later prototypes, and on the first production cars as well, could undoubtedly have been avoided if only there had been more than two development cars. Jim Dodsworth, Assistant Chief Engineer under Dick Oxley and later Project Engineer for the P5 Sportsman (Coupé) is adamant that these six cars were not enough to test and develop all the new engineering which Rover was putting into the P5. Rover did learn the lesson, of course: there were no

fewer than 16 development prototypes of the next new Rover, the P6.

THE RUN-UP TO PRODUCTION

Rover had probably started working towards a launch date of October 1958 for the P5 as soon as the third stage of the project had started in autumn 1956. However, no firm schedule appears to have been established until September 1957. Surviving documents from the Engineering Department show that the plan then was for production to build up gradually during July, August and September 1958, and for the 3-litre engine and the Borg

Warner automatic transmission to reach full production by September, one month ahead of complete vehicle production in order to ensure adequate supplies. That schedule would ensure that the car was actually announced at the Earls Court Motor Show in the same month that full production began.

Meanwhile, work had started on the Sportsman variant of the car in the early part of 1957, and Maurice Wilks was aiming to announce this alongside the standard saloon. This additional complication made the schedule incredibly tight, and Rover had to adjust its sights as early as December 1957, when the inevitable slippages led to the start date for production build-up being deferred from July to September 1958. That, of course, meant that full production would not start until December 1958 at the earliest.

Those Engineering Department documents suggest that the early part of 1958 must have been chaotic. So many areas of P5 design and development were running behind schedule that Engineering Director Robert Boyle was unable to issue a definitive production specification before March. As the production engineers needed to order tooling and plan well ahead, they were having to sign off some items for production before they had even been tested – and that in turn placed an extra load on the Drawing Office when components had to be redesigned at the last minute. The Body Drawing Office was already overloaded with work on the Sportsman body, and so further delays appeared inevitable. The engineers nevertheless did their best by reshuffling schedules, so that by the end of February 1958, production build-up was expected to begin one month earlier in August, with full production starting in November.

There was a further delay when an inter-union dispute at Pressed Steel held up delivery of the first off-tools bodies. Expected during the first week of April, these eventually reached Solihull at the beginning of May. Rover reacted swiftly, and postponed the introduction of the Sportsman model so that the Engineering and Production Departments could devote their full resources to getting the Saloon ready in time.

As there were still some elements of the car's specification which had not been settled, it was impossible to begin the design of the new assembly lines, let alone to start building them. So the Production Engineers decided to establish a temporary assembly line which would supply the Sales Department with cars for the first two or three months, and to establish the proper assembly lines later. Space for this temporary line was cleared in one of the Land Rover production areas, and work began to assemble the first bodies into complete cars during June 1958. The bodies themselves passed first through the Body Shop for checking, and were then painted in the brand-new paint plant which would come into operation for all Rover cars in the autumn.

The first cars to go down the temporary assembly line were the six production prototypes, numbered P5/1 to P5/6. Interior styling had been settled by David Bache's department as early as September 1957, but the various delays which had occurred since then meant that the production engineers had to work out how these designs could be turned into reality actually on the cars as they went down the assembly track. Work proceeded desperately slowly and the first off-tools prototype was not completed until July. Whether work continued during the traditional August factory shut-down is not clear, but the sixth and final off-tools prototype was probably not completed before September.

Robert Boyle had noted in a document dated 10th July that there was a lot of work to do on the off-tools prototypes which then

22 January 1958: the P5 Saloon (foreground) and Sportsman or Hard Top were now expected to look like this. These are actually quarter-scale models, and those gateposts are also models. The scene was set up on a board, which was put on a wall outside Denby Manor, Spencer Wilks' house at Monks Kirby. Those are real trees in the background! Note the Rover badge on the front wings and the wheel trims, inspired by Mercedes-Benz, which never went into production. The Saloon model was green, but it is not clear what colours were on the Sportsman.

The pilot-build cars

Full details of the six pilot-build 3-litres have not survived, but existing Rover records held by the British Motor Industry Heritage Trust provide information about three of them.

Car no.	Reg. no.	Reg. date	Engine no.	Remarks
P5/2	2741 NX	2.12.1959	3L7/8	Fitted with aerodynamic modifications in 1960 and used for wind-tunnel tests in July and August that year. Scrapped on 21st March, 1961.
P5/4	9922 AC	3.9.1959	3L7/5A	Fate not recorded.
P5/6	1947 AC	30.9.1958	3L7, later 6269-00296	Sold to D. Searle (Rover employee) on 22 December 1962. Subsequent disposal not recorded.

---

existed to bring them up to a satisfactory standard. One of the biggest problems was that the heavy doors put too much strain on the body pillars, and engineering reports prepared by Pressed Steel confirm that cracks began to appear at the base of the windscreen after *pavé* testing. Dan Clayton proposed the eventual solution to that, which was to prestress the door pillars so that the weight of the doors would pull them into the correct position.

Despite such serious problems, Rover went ahead with the planned launch, convinced that solutions would be found in time to get the car into the showrooms shortly after the Earls Court Motor Show that autumn. So the Rover 3-litre was introduced to the press on 22 September 1958 by Joint Managing Director George Farmer. Few, if any, of those who attended the launch can have been aware of the real reasons why Farmer had to apologize for the fact that no press demonstrators of the new car were yet available.

Assembly of the Rover P5 began as the subframe was fitted with suspension, steering gear, engine mountings and horns. The sub-assembly was then moved on a trolley to meet its engine and suspension, which were bolted in place before it was sent on its way towards the main assembly line. This picture of the powertrain sub-assembly area was taken in March 1962. Note the engines stored in racks in the background.

(Right) The subframe with its engine and transmission were next mated up to a rear axle and suspension assembly, together with a propshaft. This April 1961 shot of the assembly lines in Solihull's South Block shows how the body trim line ran alongside the drivetrain assembly line.

(Left) P5 bodies were built by Pressed Steel, initially at Linwood, then later at Swindon and after about 1968 at Cowley, Oxford. On arrival at Solihull, their undersides were treated with Bittac protective compound (which was rather primitive by later standards) and the rest of the body was primed and painted before being passed to the preparation line to receive glass and other items of door furniture. This is a two-tone bodyshell passing down the trim line in March 1962. Note how the fresh paint was protected by sheeting.

The Body Drop, part one. The prepared bodies, still lacking seats and certain items such as trim strips, grille, and headlamp surrounds, were hoisted up and across to the drivetrain assembly line, where they met up with their running-gear.

(Right) The Body Drop, part two. One man stands in the pit to hook up the rear suspension (carried on a special rocking frame during assembly), while a second lowers the body carefully into position. The body is actually supported by rods inserted into the jacking points under the sills.

(Left) With everything bolted into position, the sheeting is removed from the car before it proceeds to the body trim line for the installation of seats and carpets, and for bright trim and bumpers to be added.

(Right) Now complete, this early Mk II 3-litre Saloon has passed from the assembly line to the final preparation line, where it is being polished by an all-female team. The next stage was a road test, after which the car would be transferred to the Despatch Department for onward shipment to a Rover distributor or dealer.

The Rover Company

Today's Rover Group is not at all the same thing as the Rover Company of the 1950s, even though it shares a name and the Viking ship emblem. In those days, Rover was a small but highly respected independent company which made quality saloon cars and Land Rovers. However, after a series of mergers in the 1960s, Rover became part of the British Leyland combine. When that company needed a new name, the car division first became Austin-Rover in 1982, and then the whole company became the Rover Group in 1986. Out of all the famous marques which British Leyland had absorbed, only the Rover name was thought to have the right reputation in world markets – mainly because of the global success of the Land Rover and Range Rover.

Rover's origins go back to the last quarter of the nineteenth century, when J.K. Starley Ltd of Coventry branched out from the business of sewing-machine manufacture into the new and profitable market for bicycles. The Starley cycles sold so well that, by 1896, they had completely eclipsed the firm's original business. In that year, Starley's therefore went into voluntary liquidation, and the firm's directors formed a new company to take over the bicycle manufacturing business. Borrowing the evocative name first given to an 1884 Starley tricycle, they called it the Rover Cycle Company.

Like many other cycle manufacturers, the company turned to cars at the beginning of the new century. Its first model appeared in 1904 and, a year later, the word 'Cycle' was deleted from the company's name. Cycles (and motorcycles) continued to figure in Rover sales catalogues until 1926, but after that date the company made only cars.

However, the late 1920s were not good times for Rover. The company tried unsuccessfully to break into the big-volume small-car market and rapidly ran into financial difficulties. Not until 1933 did a change of managing director and financial director bring about a new manufacturing policy and a complete change-round in the company's fortunes.

That new policy put the quality of the Rover product above the quantities in which it was produced. The company aimed its products specifically at the well-to-do middle classes, and Rovers soon became known for being well-made, well-appointed, discreet rather than flashy, and just a little bit more sprightly than average. They became cars for the bank manager, the doctor, and the solicitor – and they sold extraordinarily well.

So well did they sell that the company was once again on a sound footing by the mid-thirties, and in 1936 was asked by the Government to take on the management of a new 'shadow' factory near Birmingham. This was one of many being built to 'shadow' or duplicate the production capacity of the country's military aircraft manufacturers at a time when war with Germany was looking increasingly likely. The major car manufacturers were asked to look after them and so to gain experience in aircraft manufacture before their own assembly plants were requisitioned for war work.

Three years after work had begun on the first factory, the Rover Company was asked to take on a second and much larger one at Solihull. Building began in 1939 just before war was declared and, by the autumn of 1940, the Solihull plant was producing its first aero engines.

After the war, it was almost inevitable that Rover should move its car production to the Solihull factory. Not only had the company's original premises in Coventry been severely damaged by Luftwaffe bombs, but also the terms of Rover's contract with the Government gave it the right to take over the shadow factories it had been running when these were no longer needed for the war effort. The Rover directors took the decision to move in principle in 1944, and the company relocated to Solihull in 1946.

The cars which followed were initially warmed-over 1940 models, but then in 1948 Rover introduced new mechanical elements in the P3 saloons, which still resembled the pre-war cars. From 1949, these were replaced by the more adventurous and much-loved P4s – familiarly known since the late fifties as 'Auntie' Rovers. And in 1948, Rover introduced the Land Rover as a stop-gap product to provide the export sales which the Government wanted. That stop-gap went on to outsell the cars by a huge margin, and its much-improved Series II models were introduced in April 1958 – just a few months before the P5s came on-stream.

2 The Mk I Models, 1958–1961

'The Rover car possesses a reputation which must be almost unique,' wrote John Bolster in *Autosport* dated 26 September 1958.

> Other makes may be faster, or have more exciting design features, but the Rover is revered throughout the world for sheer quality. It is almost taken for granted nowadays that any new car will cause its owner a good deal of trouble, but a Rover is expected to be 100 per cent reliable from the start, and to continue so indefinitely. All this is no accident, for each car goes through such a series of inspections and quality checks that any failure would be unthinkable.

The new 3-litre thus had a formidable reputation to live up to, and it was no doubt for that reason that the press were not invited to drive any of the new models when they attended the press launch at the Royal Festival Hall in London on Monday, 22 September 1958. Getting the cars assembled in time had been a minor miracle; expecting them to behave faultlessly as well might have been too much to expect. So instead, the press were invited to look around two cars on static display, one with manual transmission (car number 6259-00009) and the other an automatic. 'We shall make press cars available as soon as possible,' said George Farmer in his speech of welcome, 'so that we can have your comments and criticisms to which we attach great importance.'

The ladies and gentlemen of the press dutifully went away and recorded what they had seen. *Autosport*'s John Bolster informed his readers that 'the body appointments and equipment are of the highest Rover quality, and of course, the finish is superb. Suffice it to say that the comfort of the passengers has been studied in the most minute detail.' And he went on to conclude that, 'this car is bound to create a furore, both at home and abroad.' *The Motor* opined that 'the luggage boot is particularly roomy. As one would expect, the interior furnishing of the 3-litre Rover is beautifully carried out and sensibly planned.' *The Autocar*, meanwhile, had somehow managed to go one better.

> During a short drive in one of the early prototypes, it was evident that a considerable improvement in road-holding, when compared with the existing cars, has been achieved. There is noticeably less roll and the degree of understeer has been reduced, though there is still a trace of this characteristic. Steering is more precise, with less kick at the wheel. The driving position is very good indeed, aided by excellent all-round vision, and the relative position of the seats, pedals and steering wheel should suit the most critical of drivers.

By the time of the press launch, there were of course very few 3-litres yet in existence, and the production specification had only just been established. In fact, the very first cars

Rover 3-litre Mk I (1958–1961)

Layout

Monocoque bodyshell with front subframe bolted in place. Five- or six-seater saloon, with front engine and rear wheel drive.

Engine

Type	3L7
Block material	Cast iron
Head material	Aluminium alloy
Cylinders	Six, in line
Cooling	Water
Bore and stroke	77.8mm × 105mm
Capacity	2,995cc
Main bearings	Seven
Valves	Inlet valves in cylinder head and exhaust valves in cylinder block
Compression ratio	8.75:1 (7.5:1 for some export markets)
Carburettor	Single SU type HD6 (2in)
Max. power	115bhp gross (105bhp net) at 4,500rpm
Max. torque	164lb.ft at 1,500rpm

Transmission

Manual models	Hydraulically operated single dry plate clutch, 10in diameter
Automatic models	Torque converter

Internal gearbox ratios

Option 1 Four-speed manual

(Final drive 3.9:1)

Top	1.00:1
Third	1.37:1
Second	2.04:1
First	3.37:1
Reverse	2.96:1

Option 2 Four-speed manual with overdrive

(Final drive 4.3:1)

Overdrive	0.77:1
Top	1.00:1
Third	1.37:1
Second	2.04:1
First	3.37:1
Reverse	2.96:1

Option 3 Three-speed automatic (Borg Warner type DG)

(Final drive 3.9:1)

Top	1.00:1
Intermediate	1.43:1
First	2.30:1
Reverse	2.00:1

Suspension and steering

Front	Independent, with wishbones, laminated torsion bar springs and anti-roll bar
Rear	Semi-floating axle with progressive-rate semi-elliptic leaf springs
Steering	Burman recirculating ball type with variable ratio; power assistance optional on 1961 models
Tyres	6.70 × 15 crossply (7.10 × 15 standard for some export territories and optional elsewhere)
Wheels	Five-stud disc type
Rim width	5in

Brakes

Type	Servo-assisted drums with twin trailing shoes at the front; disc front brakes on 1961 models
Size	Drum diameter front and rear 11in
	Front disc diameter 10.75in (1961 models only)

Dimensions (in/mm)

Track, front	55/1,397
Track, rear	56/1,422
Wheelbase	110.5/2,807
Overall length	186.5/4,737
Overall width	70/1,778
Overall height	60.25/1,530
Unladen weight	3,556lb (1,613kg) (four-speed)
	3,612lb (1,638kg) (overdrive)
	3,640lb (1,651kg) (automatic)

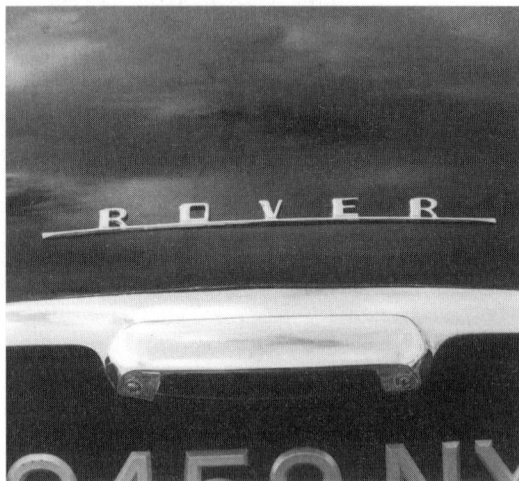

The boot lid of the Mk I bore this underlined Rover name badge. The same badge was used on the front wings of very early cars.

differed in a number of ways from the eventual production specification, although some of them were later altered to meet this production standard. As the promotional film *In the Rover Tradition* shows, the very first cars built on the temporary assembly line had bodies with a three-way colour split rather than the production two-way split, the roof being finished in the same colour as the lower body. Very early cars also carried a 'Rover' script badge ahead of the trim strip on their front wings; this was even reproduced on the first sales catalogues. There was also a unique grille badge, with the sail of the Viking ship in black on a red background (the production style had a red sail on a black background). Possibly Rover had once intended to give each model a distinctively-coloured badge, as the

(Above) *This was one of the first press pictures ever issued of the new Rover 3-litre. Taken early in August 1958, it shows one of the six off-tools prototypes and bears a registration number which belonged to another car. The 'Rover' badge on the wing looks rather wavy, and no doubt problems with fitting it on the assembly line persuaded Rover to delete it from the specification very soon after production had begun. Just visible is the 'reversed' grille badge with the black ship's sail on a red background.*

The grille badge on the Mk I cars was a metal casting with a ribbed background.

1957 and 1958 P4 105R models had a yellow and brown version of the Viking ship badge. However, the 105R was dropped when the 3-litre was introduced, and no more distinctively coloured badges were seen.

The special grille badge and the front-wing script badges are visible in the first photographs of the 3-litre issued to the press. These show a two-tone saloon which has the production style of two-way colour split, so the decision to abandon the three-way split must have been made before the these pictures were taken, which was on 6 August 1958. No production cars had been built by that date, and the car pictured is in fact one of the six off-tools prototypes built over the summer of 1958. Even the registration number of YAC 636 which it carries is false, as that number had been allocated to a P4 75 at the end of April 1958!

**Optional extras available for
the Mk I 3-litre models**

Badge bar (from January 1961)
Exhaust tail pipe finisher (from
    January 1961)
Fog lamp
Individual front seats
Laminated windscreen
Overdrive with manual transmission
    (standardized from May 1960)
Power-assisted steering (from September
    1960)
Radio (HMV or Radiomobile)
Removable division by Radford (with bench
    front seat only) (from September 1960)
Spot lamp
Two-tone paint
Tyres in 7.10 × 15 size
Wind deflector for front windows (from
    June 1960)
Wing mirrors

ON SHOW AND ON TEST

Those early press reports whetted the public's appetite for the Rover 3-litre, but the first chance most people had to set eyes on an example was at the Earls Court Motor Show which opened on 22 October. Rover had two examples there – one automatic and one manual – on its own stand (no. 161), and a two-tone green overdrive car was displayed on the Pressed Steel stand, no. 111. The Rover representatives at the Motor Show were no doubt delighted to take orders, but they could not yet guarantee delivery dates. And in fact, series production of the 3-litre did not really begin until January 1959, all the earlier cars having been more or less hand-built while production problems were ironed out.

This late start to deliveries was one reason why no road-test cars were made available to the press until the summer of 1959. The main reason, however, was that Rover was planning to replace the drum brakes on the car's front wheels with discs as soon as it could, and hoped to persuade the press to wait for the much-improved new models. The disc brakes probably arrived some time early in August 1959 – before the 1960 models came on-stream – but Raymond Mays had a set fitted to his 3-litre (CTL 17) early on and had tried them for 10,000 miles by the time he came to report very favourably on them in the 17 July 1959 issue of *Autosport*. When discs were fitted to the line-built cars, Rover made a retro-fit kit available for earlier models. Discs had been retro-fitted to 3033 AC, the road-test car lent to *The Autocar* for its report in the 21 August 1959 issue; that car had been road-registered in February and was the 269th home market car with manual transmission to be built.

The Autocar testers were extremely enthusiastic about the 3-litre. 'In the standard of comfort, appointments and silence of running it has few equals in the world, and certainly none in its price range', was their summary. The brakes were excellent, and the ride was generally good although the car seemed a little underdamped. The steering was fine at speed, although under about

---

The Royal 3-litres

Rover supplied at least two Mk I 3-litres to the Royal Household, and these were to be the first of several. In February 1961, Her Majesty the Queen took delivery of a car registered JGY 280, and in May 1961 Her Majesty the Queen Mother took delivery of a similar car registered VUL 4. Neither car was obviously different from the standard specification, although VUL 4 did carry a discreet identifying lamp at the base of its roof-mounted radio aerial.

Paint and trim colours, Rover 3-litre Mk I

1959 and 1960 model years

The first 3-litres were available in ten single-tone colours and in eight two-tone combinations. There were five upholstery colours, in each case with Charcoal Grey trimmings and carpets to match. The upholstery was Beige, Mid-blue, Red, Rush Green or Silver Grey. Wheels were coloured to suit the body paint. The standard combinations for single-tone cars were:

Body colour	Wheels
Black	Black
Dark Blue	Dark Blue
Dove Grey	Dove Grey
Dover White	Dover White
Heather Brown	Heather Brown
Light Brown	Light Brown
Light Grey	Light Grey
Rush Green	Rush Green
Shadow Green	Shadow Green
Smoke Grey	Smoke Grey

The combinations for two-tone cars were as follows (upholstery was Beige, Mid-blue, Red, Rush Green or Silver Grey):

Lower body	Upper body	Wheels
Black	Light Grey	Black
Dark Blue	Dove Grey	Dark Blue
Dover White	Light Brown	Light Brown
Dover White	Light Grey	Light Grey
Light Brown	Heather Brown	Heather Brown
Rush Green	Black	Black
Rush Green	Shadow Green	Shadow Green
Smoke Grey	Black	Black

1961 model year

The 3-litre was available in ten single-tone colours and in six two-tone combinations. There were five upholstery colours (Blue, Green, Grey, Red or Tan), in each case with Charcoal Grey trimmings and carpets to match. Wheels were painted Black. The body colours for single-tone cars were:

Black	Ivory	Medium Grey	Norse Blue	Royal Blue
Rush Green	Shadow Green	Slate Grey	Smoke Grey	Storm Grey

The combinations for two-tone cars were as follows (upholstery was Blue, Green, Grey, Red or Tan):

Lower body	Upper body
Black	Smoke Grey
Medium Grey	Ivory
Royal Blue	Slate Grey
Shadow Green	Rush Green
Storm Grey	Medium Grey
Storm Grey	Slate Grey

35mph it weighted up considerably. The testers achieved 'surprisingly good performance figures,' which included a top speed of around 97mph and a 0–60mph time of 16.2 seconds. 'Because of its size and characteristics,' the 3-litre was 'better suited to be chauffeur-driven than any of its predecessors.' There were drawbacks, however. There was axle hop under hard acceleration, the notchy gear lever was 'far from pleasant' – a fairly damning criticism in the days when road-test reports habitually pulled punches – and the Dunlop Gold Seal tyres squealed embarrassingly during even gentle cornering.

The Motor test of an automatic 3-litre in the issue dated 6 July 1960 pulled its punches magnificently. 'Neither very high performance nor extreme economy of petrol has been a primary objective of the design' was the way in which the testers explained that the car was slow and thirsty, while

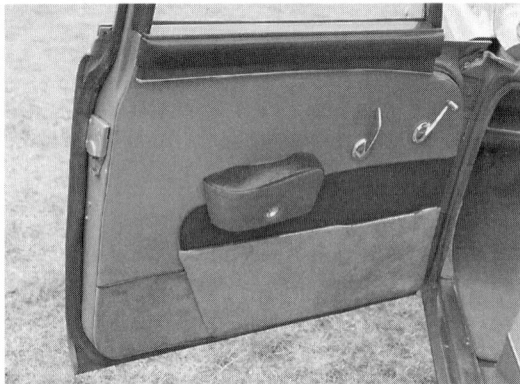

(Left) The Mk I style of armrest had a push-button in the centre to allow vertical adjustment.

Charcoal Grey panels made a contrast with the main interior trim colour. It seemed like the height of fashion in the late fifties, but dated very quickly. This is an automatic model, and the '2nd Gear Hold' switch can be seen next to the cigarette lighter under the dash – not the most accessible of places for the driver! This is an early car, with leathercloth around the door tops and just a slim strip of wood beneath the padded roll under the window. Jim Dodsworth, Project Engineer on the Sportsman (Coupé) model, says that the characteristic lip on top of the instrument binnacle was developed to prevent reflections in the Sportsman's more raked and curved windscreen, and that it was adopted for the Saloon models as well.

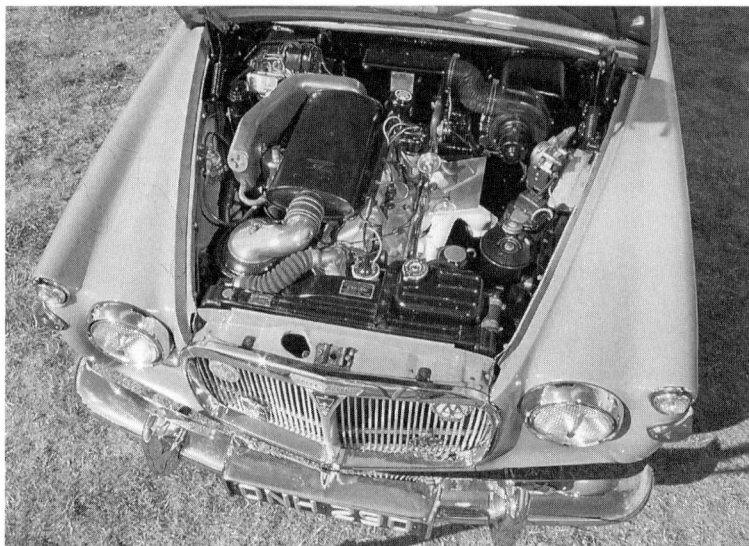

The original 3-litre engine had an oil-bath air cleaner and developed 115bhp – adequate, but not enough to give the car 100mph performance.

'there are aspects of its character which remind a driver that this model has not yet enjoyed as many years of detailed development as have the smaller Rovers' meant that the car was underdeveloped! The magazine's staff did not think the car was up to the standards set by the P4 models:

> In inevitable comparison with the smoothness and silence which other Rover models have acquired as a result of development work extending over many years, the 3-litre seems as yet to represent a slight retrogression, although it has undoubtedly been improved to a very great extent since its first announcement in September 1958.
>
> (Or, translated: the first cars were terrible and even this one is not as good as we have come to expect of Rover.)

It may well have been that there was a fault with the test car, 6519 NX (6300-00820), because the testers complained that the disc brakes juddered and pulled to one side. Other complaints were more understandable, though. The testers thought the

intermediate-speed hold control was awkwardly placed, that the steering was heavy at parking speeds and that the suspension allowed too much vertical movement of the body over poor surfaces. The wipers did not sweep the whole windscreen, and front end shake over poor surfaces was inexcusable in a car of these pretensions. On the credit side, however, ventilation was good and the car was very roomy.

Autosport's regular road test correspondent, John Bolster, tested out a disc-braked overdrive example for that magazine's issue dated 16 September 1960. He concluded that:

> The Rover 3-litre is a car for the man who cannot bear anything cheap and flashy. He buys his car knowing that he will be living with it for a long time, and he chooses it because it will give him smooth, silent travel in an atmosphere that does not affront his good taste.

He thought the 3-litre was 'every inch a Rover', and his only real criticism was that

Mk I 3-litre: commission numbers and production figures

1959 Mk I	6259-00001 to -01584	Manual, home market	(1,584)
	6269-00001 to -00305	Manual, export RHD	(305)
	6289-00001 to -00186	Manual, export LHD	(186)
	6309-00001 to -00768	Auto, home market	(768)
	6319-00001 to -00158	Auto, export RHD	(158)
	6339-00001 to -00159	Auto, export LHD	(159)
			Total: 3,160
1959 Mk I Station Wagon (prototype only)	6309-70001 (one only)	Auto, home market (built late 1959)	(1)
			Total: 1
1960 Mk I	6250-00001 to -03061	Manual, home market	(3,061)
	6260-00001 to -00971	Manual, export RHD	(971)
	6280-00001 to -00368	Manual, export LHD	(368)
	6300-00001 to -02236	Auto, home market	(2,236)
	6310-00001 to -00511	Auto, export RHD	(511)
	6330-00001 to -00599	Auto, export LHD	(599)
			Total: 7,746
1961 Mk I	6251-00001 to -01661	Manual, home market	(1,661)
	6261-00001 to -00376	Manual, export RHD	(376)
	6271-00001 to -00006	Manual, CKD RHD	(6)
	6281-00001 to -00212	Manual, export LHD	(212)
	6301-00001 to -01532	Auto, home market	(1,532)
	6311-00001 to -00391	Auto, export RHD	(391)
	6331-00001 to -00166	Auto, export LHD	(166)
			Total: 4,344
			Total: 15,251

Rovers always came well equipped, and this is the set of major tools clipped to the inside of the boot: a starting handle, wheelbrace, jack and tyre pump.

This artist's impression – another early publicity picture – shows the wishbone front suspension with the gaitered torsion-bar spring running fore-and-aft and the anti-roll bar ahead of the wheel centre line.

the engine went through a vibration period at the top end of its rev range. As compared to earlier Rovers, however,

there is no spectacular increase in performance. Instead the extra power of the bigger engine is used to propel a wider and more luxurious body than that of any previous Rover.

Motor Sport did not get to test a 3-litre until April 1961, when writer Bill Boddy had the use of an overdrive car equipped with power steering. This was 2902 UE, which was actually the first 1961-model home market manual car (number 6251-0001). Boddy believed that,

It takes a couple of hundred miles or so before the true worth, and a few minor shortcomings, of the 3-litre are revealed. Its weak points are flexible suspension giving too much up-and-down motion and some roll on corners, while the unitary construction is not entirely free from shake over bad roads. Otherwise, this is a

beautifully appointed and fully-equipped car, capable of truly quiet and effortless cruising in the very high top and overdrive gears. It competes with a certain twin-cam Coventry-built car in price but offers more spacious accommodation and will undoubtedly appeal to connoisseurs of good cars, particularly those who have had a long line of Rovers.

Boddy was particularly impressed by the car's spaciousness, and he found its performance 'ample', but noted that 'acceleration is smooth rather than vivid, although the smooth silent flow of power can be deceptive.' He thought the gear lever left something to be desired and that it was too easy to get reverse instead of first, pointed out that the positions of the minor switches needed to be learned, and complained that the wipers left unswept areas at the sides of the wrap-around windscreen. The car was equipped with Avon tyres, which Boddy pumped up above the recommended pressures, and which gave silent running in all conditions.

The Chapron Drophead

It is no longer clear why Rover asked the French coachbuilder Henri Chapron to turn a 3-litre saloon into a drophead at the beginning of 1962. The idea might have come from Maurice Wilks, who had been behind Rover's look at a convertible P4 in the first half of the fifties, and whose boyish enthusiasm for new projects was almost legendary. Certainly, the contract drawn up for the car's construction in January 1962 specified that when the car was returned to Rover, it should be in a fit condition to be sold to a member of the company for his personal use.

In the early sixties, Chapron was France's leading coachbuilder, and his cabriolet conversion of the Citroën DS was widely recognized as one of the most attractive large open cars around. Mulliner's of Birmingham and Salmons-Tickford, whom Rover had previously commissioned to build special bodywork, were no longer in a position to take such commissions, and this may explain why a foreign coachbuilder was selected to work on the 3-litre. Mulliner's had been bought by Triumph in the mid-fifties, while Tickford had become part of the Aston Martin-Lagonda group.

The approach to Chapron was made through Rover's French importers, Société Franco-Britannic Autos whose head office was in the same Paris suburb of Levallois-Perret as Chapron's studio. A contract was drawn up between the Rover Company and Franco-Britannic, under which Rover would lend the French company a 3-litre saloon so that Franco-Britannic could investigate whether it could be converted into a drophead. Franco-Britannic then subcontracted the work to Chapron.

The car supplied to Franco-Britannic was a 1959 right-hand drive Mk I with manual transmission, number 6259-01444. It had been registered by Solihull as 9823 AC on 10 September 1959, and had probably been used as a development vehicle by the Experimental Department. Chapron turned it into an extremely elegant drophead in time to display it on his company's stand at the Paris Motor Show in the autumn of 1962.

Chapron design number 7472 retained the standard Rover front panels, windscreen and boot lid, but the saloon body was heavily modified in the centre section to give just two doors. The B-pillar was moved backwards and longer front doors were built to improve access to the rear seat. Meanwhile, the roof was cut off just behind the windscreen, and Chapron added a smart convertible top which folded down almost flush with the rear deck. As on Chapron's Citroën design, there were no windows in the rear body sides, and the top looked sleek and neat when erected. Its large blind quarters must nevertheless have made rearward visibility a problem for the driver and would have restricted the view out for the rear seat passengers. Behind the hood well, the standard chromed petrol filler cover was removed – perhaps because the tank had been altered to allow room for the folding hood behind the seats – and was replaced by a rectangular locking cover panel. The rear seat was also made narrower to accommodate the folding hood mechanism, and its backrest may have been made slightly more upright than standard.

Chapron's file for the car shows that the convertible top was made of grey cloth. The seats and hood cover were in dark grey and there were contrasting red carpets. The bodywork was finished in 'English paint', which surviving black-and-white photographs suggest might have been a silver metallic. Finishing touches were a modified trim strip on the flanks with the name 'Henri Chapron' discreetly added to the front wing in chromed capitals, and a bright finisher strip below the doors and continuing along the wing bottoms.

The car was re-registered as 1601 MY 75 in Paris after conversion. Although the terms of the contract with Franco-Britannic specified that the car would be returned to Rover for evaluation, those who would most likely have seen it at Solihull do not remember it at all. Its subsequent history is unknown; one story suggests that the car was later sold in Switzerland, although there may be some confusion here with the later Graber 3-litre cabriolet. The contract between Rover and Franco-Britannic specifically forbade the French company to go ahead with series production of the conversion unless Rover gave permission, and Chapron was not asked to build any more drophead 3-litres. That was a pity, as his car was undoubtedly the most attractive of the drophead conversions made over the next few years.

(Left) *The Chapron convertible commissioned by Rover in 1962 and made from a 1959 3-litre was a particularly elegant car. What a pity it has now been lost.*

Chapron re-upholstered his 3-litre convertible in a style completely in keeping with the car's luxury pretensions. Note how the hood folded away almost flush with the body, and how the rear seat lost some width to the hood's folding mechanism.

RIVALS

The 3-litre represented a move into a market sector which Rover had not previously tackled. The company's existing P4 saloons were well established as appealing to the wealthy professional man, but with the more expensive and better-equipped P5 Rover hoped to be able to sell in the slightly more rarefied air of the luxury market, just above that. If the P4 was the car for the solicitor or bank manager, the P5 was intended to be the car for company directors.

Right at the top end of that market were the products of Rolls-Royce, and some way further down came struggling small-volume makers like Armstrong-Siddeley and Daimler, both of whom would soon be extinct (although the Daimler name survived after

the 1960 Jaguar takeover). Rather below that came Jaguar, priced not too far above the existing Rover saloons and offering a blend of luxury and high performance in its larger saloons. Rover's plan was to offer a car in the Daimler or Armstrong-Siddeley mould, but to sell it in relatively high volumes and so keep its price just below that of the big Jaguars.

The pricing of the first 3-litres on the home market made quite clear what Rover was trying to do. Before purchase tax and extras were added, the 3-litre with four-speed manual transmission cost £1,175 and its automatic equivalent £1,280. These prices neatly bracketed the overdrive version of the Mk VIII Jaguar at £1,264. The only other large-engined saloon in the same price bracket was the smaller 3.4-litre Jaguar saloon (£1,242 with automatic transmission). For comparison, the cheapest Armstrong Siddeley Star Sapphire was well over £1,700, the cheapest Daimler well over £1,600 and the Rolls-Royce Silver Cloud cost nearly £3,800.

The position had changed hardly at all by the time of the 1959 Earls Court Motor Show. Jaguar had moved its big saloons (the new Mk IX had actually been announced alongside the run-out Mk VIIIs at the 1958 show) into a higher price bracket and thus removed some potential competition for the 3-litre. A price increase in July 1959 had put the 3-litre up to £1,210 in manual form and £1,315 as an automatic, but once again those prices neatly bracketed the small Jaguar, which was revitalized as the Mk 2 model at the 1959 show and now cost £1,282 in automatic form. The Mercedes-Benz 190 was in the same price bracket at £1,250, but it was no luxury car and found very few buyers, despite the excellence of its engineering and build. During the 1960 season, sales of the Rover 3-litre more than doubled the first season's totals.

It is at least arguable that the Rover 3-litre had created a new market for itself, and that for the first two years of its existence, it stood alone in that market. However, the competition soon became very much stiffer. BMC had announced its new big saloons at the 1959 Motor Show, and among them were the Farina-styled 3-litre Austin A99 and Wolseley 6/99. Not long after that, BMC decided to tackle the Rover by introducing a more luxuriously finished derivative of these called the Austin Princess 3-litre, which in May 1960 was renamed a Vanden Plas Princess 3-litre. It was not as fast as the Rover, and nor did it handle as well, but it was priced very competitively indeed at a couple of hundred pounds below the Solihull car. In automatic form, it cost just £1,035 before purchase tax by the time of the 1960 Earls Court Show.

That Motor Show saw yet another manufacturer attempt to muscle in on the market which Rover had created with its 3-litre. This time it was Humber, whose Super Snipe had appeared at the 1958 Earls Court Show as a rather underpowered 2.6-litre but now took on a 3-litre engine with 129bhp. At £1,050 before purchase tax, it was astonishingly good value even though it did not undercut the Vanden Plas Princess. The Rover 3-litre, now up to £1,258 in overdrive form (overdrive had been standardized in May 1960) but still £1,315 as an automatic, appeared way overpriced in the face of this BMC and Humber competition, and sales during the 1961 season dived to around 55% of their 1960-season levels. It was clearly time for Rover to make some drastic changes.

PRODUCTION CHANGES

The 3-litre had more than its fair share of teething troubles during the first two years of production. Most new cars have a few, as problems which did not become apparent during the development period show up in service, but the number of changes Rover made to the cars during the 1959 and 1960 seasons suggests that lessons were being learned very fast at Solihull.

There were, as already explained, difficulties in getting the cars into production at all during the last few months of 1958. In the first few months of full production during 1959, the brakes had to be modified to prevent squeal, the front bumper wraparound had to be lengthened to improve protection against parking knocks, a heavier grade of oil had to be specified for the manual gearbox and the engine mountings on manual-transmission cars had to be changed to prevent vibration. These were certainly problems which should have been identified during pre-production development.

By this stage, Chris Goode had moved on and Dan Clayton had become P5 Project

Rover 3-litre Mk I: production changes

Note: All dates given for changes where no chassis number is quoted must be treated as approximate. The dates are taken from issues of *Rover Service Newsletter*, which usually reported changes between one and four months after they had actually been made on the assembly lines. Dates given for changes where a chassis number is quoted are exact for home market models and reflect the date into Despatch (i.e. the date the car was transferred from the assembly lines to the Despatch Department) given in records held by the British Motor Industry Heritage Trust.

Date	Home, manual	Home, auto	Other	Remarks
September 1958	3-litre introduced			
January 1959				Recommended fluid change interval for automatic transmission reduced from 24,000 to 12,000 miles.
February 1959	6259-00416	6309-00150	6269-00031 6289-00016 6319-00004 6339-00006	Anti-squeal ring added to front brake drums.
March 1959	6259-00503	6309-00181	6269-00035 6289-00017 6319-00004 6339-00008	Increased wraparound for front bumper ends; retro-fit possible.
April 1959	6259-00710	6309-00283	6269-00101 6289-00062 6319-00020 6339-00036	Rubber-mounted propshaft centre support added; retro-fit possible.
				Red plastic cover with gold Rover crest available for Owner's Instruction Book.
May 1959				Champion N5 spark plugs replaced Lodge CLNH type on 8.75:1 CR engines; 7.5:1 CR engines retained Lodge CLNH plugs but Champion N8 also now approved.
				Nylon filter replaced rubber bodied type in brake fluid reservoir.
June 1959				40SAE oil recommended in place of 20SAE in manual gearbox.
				Optional cylinder block immersion heater introduced, with either 100–125 volt or 200–250 volt rating; socket fitted to parcel shelf.
July 1959	6259-01121		6269-00197 6289-00129	Softer rubber for rear engine mounting on manual-transmission cars.

	6259-01204	6309-00588	6269-00212 6289-00139 6319-00107 6339-00110	Redesigned steering column mounting bracket; retro-fit possible.
	6259-01263	6309-00625	6269-00225 6289-00150 6319-00117 6339-00121	Wiper motor repositioned with shorter cable run to improve efficiency; brake fluid reservoir and pipes on LHD models moved to suit.
	6259-01286	6309-00628	6269-00234 6289-00155 6319-00125 6339-00135	Rubber fuel filler hose replaced by metal tube with rubber hoses at each end, to improve filling and venting and minimise evaporation.
August 1959	6259-01307		6269-00176 6289-00121	Spring balance over-centre linkage introduced, to give lighter clutch pedal operation; retro-fit conversion kit available.
	6259-01419	6309-00679	6269-00068 6289-00172 6319-00138 6339-00137	Door glass support bracket introduced, to prevent glass rattling when lowered; retro-fit possible.
				Modified steady post and link pin in front brakes (date approximate only).
September 1959	6259-01462	6309-00711	6269-00275 6289-00176 6319-00143 6339-00143	Fuse covering fog lamps and wipers changed from 35 amp to 50 amp type.

1960 model-year

Front disc brakes standardized.

Modified gear selector pivot shaft to improve access to torque converter checking point on LHD automatic cars.

Lucas C45 PV6 dynamo with RB310 regulator replaced CV45 PV5 type to raise output from 22 to 25 amps.

Lucas 26164A starter fitted, with higher torque.

Large notched drain plug for overdrive unit replaced hexagon-head plug.

Tyre pressures for 6.70×15 Michelin X replacements recommended as 26psi (F) and 30psi (R).

October 1959	6250-00070	6300-00021	6260-00018 6280-00010 6310-00004 6330-00017	Improved wiper switch to prevent internal shorting.
	6250-00379	6300-00163	6260-00120 6280-00047 6310-00041 6330-00095	Modified hose layout for brake servo.
	6250-00379	6300-00163	6260-00120 6280-00047 6310-00093 6330-00226	Softer rubber for rear spring mountings, to improve ride.
				Manual gearbox mainshaft modified to prevent jumping out of top and 2nd gears; synchronizing clutch for 3rd and 4th modified to suit; larger mainshaft nut.
November 1959	6250-00456	6300-00185	6260-00148 6280-00060 6310-00054 6330-00124	Two pipes with short interconnecting hose replaced one-piece pipe between brake servo and reservoir tank; retro-fit possible.
		6300-00190	6310-00055 6330-00125	Modified kick-down spring on automatic cars.
				Front toe-in changed from ⅛in to ¹⁄₁₆in.
				Carburettor modified to improve slow running.
December 1959	6250-00864	6300-00403	6260-00301 6280-00106 6310-00106 6330-00288	Ornamental plate on tool tray front deleted, to improve passenger knee-room.
January 1960	6250-00744	6300-00749	6260-00432 6280-00176 6310-00193 6330-00444	Electric screen wash replaced foot-operated type; retro-fit possible.
				Lodge HBLN spark plug approved as service replacement on 8.75:1 CR engines.
				Automatic transmission modified to give firmer up-change (from box no. R5B-2402).
				Exhaust heat shields not to be made available; drawings for hand manufacture available to dealers.

February 1960	6250-00980	6300-00460	6260-00333	First stage of revised body trim, affecting: roof, windscreen, scuttle, cant rail, D-post, rear lower quarters and BC posts. Rear parcel shelf and rear window rubber changed; oblong courtesy lights above doors; upholstery pleating now front to rear; driving mirror changed; metal coat hooks added behind BC pillars.
			6280-00111	
			6310-00116	
			6330-00311	
	6250-01036	6300-00524	6260-00371	Expansion tank added to petrol tank for better breathing and fume control.
			6280-00122	
			6310-00129	
			6330-00351	
				Improved dust-sealing with new compression-type door seals, mud excluders on rear door over wheelarch, and rubber grommet in door check strap lubrication holes.
March 1960	6250-01353	6300-00749	6260-00432	Second stage of revised body trim, affecting: facia frame and mouldings, fresh air controls (butterfly-type air vents introduced), glove box lid and mouldings, door casings and door mouldings. (N.B. 6250-01351 also had the revised trim, and 6250-01365 had the old trim.)
			6280-00176	
			6310-00193	
			6330-00444	
	6250-01429	6300-00832	6260-00448	Modified oil filter, with fixing bolt moved from bottom to top.
			6280-00193	
			6310-00213	
			6330-00476	
April 1960	6250-01474	6300-00856	6260-00459	Improved fuel pump, type AUA 82.
			6280-00202	
			6310-00221	
			6330-00486	
	6250-01837	6300-01043	6260-00551	Steel spacer replaced diecast Mazak spacer for recuperating seal in brake master cylinder.
			6280-00255	
			6310-00265	
			6330-00529	
	6250-01885	6300-01091	6260-00578	Modified exhaust mounting at bell-housing, to eliminate tendency to howl.
			6280-00271	
			6310-00273	
			6330-00534	
May 1960				Overdrive standardized with manual transmission from 1st May.

June 1960	6250-02318	6300-01517	6260-00741 6280-00327 6310-00415 6330-00591	Thicker steering drop arm with raised ball centre, to improve steering geometry.
				O-ring and (later) felt seal introduced for rear wheel bearing housing.
				Side-entry distributor replaced top-entry type.
				Improved design of crankshaft oil retainer.
				Bolt-on swivelling wind deflector available for front door windows.
undated, before June 1960				Kit introduced to stiffen cushion on bucketseats.
				Kit introduced to convert bench seat to seat to buckets.
				Kit introduced to to modify rear parcel tray and prevent trim from buckling.
				Kit introduced to fit distribution baffle to heater to improve warm air distribution.
July 1960				⁷⁄₁₆in drain plug replaced ⅜in plug in torque converter.
August 1960				Larger-bore bush for distributor drive-shaft.
				Longer dowels for rear main bearing oil seal.
September 1960	6250-02964	6300-02166	6260-00951 6280-00365 6310-00503 6330-00600	Door glass support bracket pad now without rivets.
	6250-02998	6300-02185	6260-00955 6280-00367 6310-00507 6330-00600	Modified steering coupling flange and locating plate.

1961 model-year

				Heater and blower output increased by 25 per cent.
				Improved door rubbers for better sealing.
				Automatic transmission now had second-speed start (from first 1961-season engines).

November 1960				Handbrake-on warning light for automatic cars.
				Self-locking nuts in steering coupling disc.
				Stainless-steel body trim mouldings replaced chrome-plated brass.
				Power-assisted steering option introduced.
				Radford division option introduced.
				Reinforcing sleeve added to fuel tank overflow hose.
				Clutch plate decelerator fitted, from gearbox 6250-03710.
				Magnetic drain plug for overdrive unit.
				Propshaft UJs now had pre-packed and fully-sealed bearings.
				Improved underbonnet sound insulation; bonnet panel modified to suit.
December 1960	6251-00707	6301-00439	6261-00156 6281-00054 6311-00207 6331-00018	New boot lid lock.
				Camshaft commonized with P4 100, from engines 6251-00012, 6261-00076, 6271-00011, 6281-00005, 6301-00139 and 6311-00050.
				Phosphor-bronze oil pump drive shaft gear replaced steel type, from engines 6251-00018, 6261-00170, 6271-00045, 6281-00032, 6301-00139 and 6311-00137.
January 1961				Badge bar available.
				Exhaust tail pipe finisher available.
				Insulation sleeves added to terminals on certain dash switches.
February 1961	6251-00889	6301-00602	6261-00223 6281-00089 6311-00306 6331-00041	Front hub modified to prevent oil seal working loose.

Date				
				Counterbore added to hub end of camshaft; thread depth increased.
				Modified water pump spindle, from engines 6251-00041, 6261-00520, 6271-00045, 6281-00032, 6301-00347 and 6311-00137.
				Stronger reverse stop spring, from gearbox 6251-00708.
March 1961		6301-00523	6311-00296 6331-00025	Strap support for compensator shaft on automatic transmission replaced bearing support.
	6251-00989	6301-00707	6261-00249 6281-00109 6311-00329 6331-00059	Centre pressing of road wheels modified to improve wheel seating and eliminate rear brake judder.
				Minor internal modification to automatic transmission and modified upchange speeds.
				Fuel tank overflow pipe modified: long copper pipe and short hose replaced short copper pipe and long polyurethane hose; felt pad added between pipe and tank.
April 1961	6251-01044	6301-00767	6261-00261 6281-00110 6311-00340 6331-00067	Improved battery box seals.
May 1961	6251-01046	6301-00840	6261-00263 6281-00110 6311-00353 6331-00078	Five-point engine mountings introduced; car numbers given are approximate only.
	6251-01046	6301-00840	6261-00263 6281-00110 6311-00353 6331-00078	Lock for glove box lid recessed further to increase clearance between handle and lock; ashtrays changed: two new-style ashtrays per car replaced one old-style.
		6301-00840	6311-00353 6331-00078	Automatic transmission had modified clutch linings to give smoother shifts; modified transmissions identified by a P in the serial number.
	6251-01053	6301-00841	6261-00263 6281-00110 6311-00354 6331-00079	Modified front exhaust mounting bracket, to give two-plane adjustment.

	6251-01171	6301-00960	6261-00280 6281-00146 6311-00364 6331-00093	Improved interior lamps, now with festoon type bulbs.
				Kit available with modified engine mounting brackets and rubbers to prevent lateral shake at 55–75mph.
				Shallower radiator filler cap and new header tank, to commonize caps with P4 models.
June 1961	6251-01328	6301-01088	6261-00316 6281-00179 6311-00377 6331-00111	Foam rubber insert added to fuel filler seal.
				Painted pointer on automatic gear selector quadrant replaced by chromed ring.
				Kit introduced to extend steering column by 2½in.
				New door lock striker plate with four-point fixing instead of three-point fixing.
				Locking spring added to spare wheel housing operating screw.
July 1961				7.10 × 15 tyres no longer available as options, but still standard for Austria and Switzerland.
August 1961				Conrods had rolled-thread bolts instead of bolts with machine-cut threads.
				Modified gearbox selector shaft to prevent noise when selecting 2nd gear, from gearbox 6251-01418.
September 1961				Bulbs repositioned in rear number plate light housing.

Engineer. He had joined the P5 Project team as one of a small number of Technical Assistants in 1957, and had then been promoted to the vacant position of Assistant Project Engineer. He remembers in particular the inadequacies of the original braking system:

We'd decided the sooner we got off drum brakes, the better. Before the vehicle was announced we were on to trying disc brakes. They were fairly young ... there was no great track record on disc brakes and there was a lot of concern about excessive pad wear.

Someone in the Engineering Department – probably Robert Boyle – had conceived a plan to send a 3-litre with disc brakes out to Kenya for testing on that country's dirt roads by Cooper Motor Corporation, who were then

The original squat bumper brackets were quite unlike the later rod type.

Rover's East African distributors. This was a major departure for a motor manufacturer: overseas testing in arduous conditions simply did not figure in the normal development procedures of the time, and even Land Rovers were not subjected to dirt-road testing. Dan Clayton went out with the car to supervise the Kenyan test, and remembers:

The thing that didn't give us any trouble was the disc brakes, funnily enough!

But the problems we came across were almost without number! It was a completely different environment, and everything was strange compared to the testing that we'd put the vehicles through. This was a new experience for us and for the car, and it suffered in all sorts of ways. The amount of information I sent back and brought back with me caused quite a stir.

The disc front brakes were the only major change introduced on the 1960 model-year cars in September 1959, and they made a huge difference to the car, which could now be driven fast with confidence. Even the colour schemes remained as they had been for the 1959 models, but both the dynamo and the starter motor were uprated, probably because starting problems in service had left owners with flat batteries. Nevertheless, several modifications were still to come, and during the 1960 season Rover lightened the clutch, stopped the door windows rattling, stopped petrol fumes seeping into the boot, softened the ride and prevented the manual

The Kenyan 'safaris' early in 1960 taught Rover a lot about the 3-litre. This is one of the test cars, 1865 NX, with its crew at the Equator.

The optional division made by Harold Radford was available only for use with the bench front seat at this stage. It was trimmed to match the upholstery and was easily removable.

cars from jumping out of gear. The list went on; it was not a pretty picture.

Customer feedback must have been the main reason why the interior trim was radically changed in mid-season. David Bache's original design must have proved too modern for conservative Rover owners, because the modified trim made the 3-litre's interior look much more like that of the much older P4 model. The new trim was introduced in two stages during February and March 1960, and its main features were wood trim around the door windows in place of leathercloth, front-to-rear seat pleating instead of side-to-side pleats, more discreet fresh-air vents on the facia and new door trims without the contrasting charcoal-coloured panels. The result was undoubtedly a change for the better, giving the interior a cleaner and less cluttered appearance while giving more prominence to the polished wood trim. Rover owners always have liked their wood and leather: just over a decade and a half later, the Rover SD1 disappointed many customers when it was announced without either in its passenger compartment. Rover eventually bowed to popular demand by introducing both after the car had been in production for six years.

Meanwhile, Rover had been back to Kenya with some 3-litres. Dan Clayton had suggested that the design engineers and development engineers would learn a great deal of benefit to Rover products by going through the same experience as he had been through on that country's dirt roads and in its hot climate. So between 18 January and 14 April 1960, five teams of engineers went out to Nairobi to drive some 3-litres around a long-distance circuit prepared with the assistance of Cooper Motor Corporation. Those who went included Clayton himself, Gordon Bashford from the Advanced Design Department and the representatives of some of Rover's suppliers, notably from Pressed Steel.

The exercise proved enormously beneficial to all concerned, and generated a large number of reports. The Pressed Steel representatives, for example, were mainly interested in observing how the structure of the bodyshell which their company supplied stood up to the punishing schedule. For them,

The objective was to drive the car as hard as road conditions would permit, to discover as many potential faults as possible and to compare the results obtained under conditions of high altitude and temperature with those obtained from a similar vehicle which is undergoing extensive testing at Armoured Vehicle Establishment, Chobham.

Their report listed just a few of the many weaknesses which showed up in the cars used on the test: weaknesses in the spare wheel pan mountings, the rear spring shackle brackets, the body pillars and the boot lock (which a Rover representative told them was never satisfactory from the beginning!). The dust sealing of the boot also proved to be inadequate. After that Kenyan 'safari' was over, everyone involved went away to do a lot of hard thinking. Not much happened in time to affect the 1961-season 3-litres which were due to be announced that autumn, but in the longer term, Rover learned a great deal of value from the exercise. And it led on in due course to the appearance of the 3-litre in international long-distance rallies, as Chapter 9 reveals.

Even so, the 1961 cars did incorporate a number of other changes, and in fact the minutes of the Rover Board record that the 1961-season modifications cost a not inconsiderable £100,000. An underbonnet insulation panel had been added to reduce engine noise, and automatic models now started in second rather than first gear in order to give smoother progress. This arrangement, of course, also slowed acceleration, but it was possible to kick the transmission down to first gear for a more vigorous getaway. On manual cars, all 1961 models came with overdrive, which had actually been standardized in May 1960.

The cause of greater convenience was served by an electric screenwash in place of the earlier foot-operated pump, but probably the most important change of all was that power-assisted steering was now available. The system Rover had chosen was a brand-new one made by Hydrosteer and operated by a pump fitted to the back of the dynamo and thus driven by a belt from the engine crankshaft pulley. The absence of power-assisted steering had been a major hindrance to sales of the 3-litre in the USA, and its introduction – as an extra-cost option – helped the car live up to luxury-market expectations at home as well, even though the non-assisted steering was in fact perfectly acceptable in most conditions.

Nevertheless, the new season's cars were most obviously distinguished by a new selection of paint and trim colours. To simplify production, Rover had also decided to paint all wheel discs in black instead of in the variety of colours used on 1959 and 1960 models to make the wheels tone in with the body colour. The only other cosmetic change for 1961 was less noticeable, but important for the durability of the cars: the side trim strips were changed to stainless steel from the chrome-plated brass which tended to pit and corrode quickly on the earlier 3-litres.

With the 1961 cars, Rover had very nearly got the 3-litre right, although there were still shortcomings. Acceleration was one of them, and Solihull's engineers were already working on improving both the torque and the power of the six-cylinder engine. However, before this engine was ready for introduction, there was the question of engine shake to be addressed once again. The softer rear mounting introduced during the 1959 season on manual cars had helped to a certain extent, but it had not cured the problem entirely. So in May 1961, the original three-point engine mounting arrangement gave way to a more robust five-point type, a modification which cost Rover £27,000 to introduce – once again, a not inconsiderable sum.

THE 3-LITRE IN THE USA

Rover's relationship with the USA was always a slightly difficult one. The Rover Company was a small one and could simply not afford to establish a chain of dealerships right across the USA, and so in the early fifties it 'piggy-backed' on the much larger Rootes Group organization across the Atlantic.

However, Rover embarked on a new strategy during 1959. In the first part of the year, it established a new subsidiary, the Rover Motor Company of North America, with headquarters in Toronto, Canada. The US branch was established at Long Island City, New York, and parts depots were established in Toronto, Vancouver, New York and San Francisco. A further sales office was also established in Montreal. RMCNA President H. Gordon Munro announced confidently during 1959 that:

> The Rover Company has launched a vigorous campaign, from the far north to the Rio Grande and from the Atlantic to Pacific, to capture an important share of the North American market for quality cars and for utility vehicles.

Sadly, neither the Land Rovers nor the P4 saloons would ever achieve anything like

This publicity picture for the 1960 model year shows a 3-litre in the USA wearing Californian plates. Note the special headlamps (cars for most other markets had the Lucas 'tripod' type) and the whitewall tyres. The contrast in styles between the Rover and the befinned American cars in the background is enormous.

52

the important share Munro envisaged, and the introduction of the P5 towards the end of 1959 made hardly any difference at all. Nevertheless, RMCNA did do its best to gather publicity for the car, and lent a four-speed example to *Car Life* magazine and an automatic example to *Road and Track* (whose California-based testers were a little miffed at having to travel to New York to collect the car!). Overdrive models were of course available as well.

It must be said that Rover's comprehension of the US market in the late fifties was limited. The company's Directors knew that US customers were generally receptive to European imports at the time, and they felt sure that Americans would appreciate the traditional British qualities of Rover saloons – they had, after all, taken to Rolls-Royce and Jaguar saloons partly because of those qualities. However, Rover did not expect to alter its products to any great extent to suit American motoring habits or requirements. So it was that the first 3-litres which went to the USA had wide-band whitewall tyres as their sole concession to American tastes.

Road and Track found the acceleration of the 3-litre automatic rather underwhelming in its December 1959 road test, and commented on the car's body roll and tyre squeal during cornering. However, the instrument layout and brakes (by now with front discs) came in for high praise, and refinement was also a strong point. 'With the exception of one car (we don't mention the name, but it's synonymous with quality) this is the most refined automobile we've ever driven,' the magazine's testers stated. The price was high, though: at $4,995 plus delivery costs, the 3-litre cost as much as an entry-level Cadillac and more than twice as much as the cruder but similarly-sized 'compact' Ford Falcon with its optional 145bhp 3.6-litre straight-six engine.

Car Life also wondered about the high cost of the 3-litre when it reported on a non-overdrive manual model in its March 1960 issue. The tester thought the gearbox should at least have synchromesh on all four gears in a car costing nearly $5,000, and would have liked power-adjustable front seats (which were, of course, unheard of on any European car at that time). He pointed out that in size the 3-litre was comparable to the latest American 'compacts' like the Ford Falcon (109.5in wheelbase, around $2,000), Rambler (the Six had a 108in wheelbase and cost around $2,500) and Studebaker Lark (108.5in wheelbase and priced between $2,100 and $2,500).

The tester also thought the Rover needed power-assisted steering for low-speed work, and criticized the steering for being slow and imprecise even on the move. The ride quality was 'disappointing', and the louvres over the windows whistled at speeds above 45mph. Nevertheless, 'little effort has been spared to make the Rover's interior as quiet, comfortable and elegant as a London club,' and the build quality left little doubt that the car would last a long time. It was, the tester summarized, 'a sedate and solid

The fuel filler was concealed under this chromed cap. The centre section concealed a keylock.

luxury package almost completely lacking in sporting characteristics.'

When *Sports Cars Illustrated* tested a manual-transmission car for its May 1960 issue, the model was equipped with the all-drum braking system which had gone out of production the previous summer. Newer cars, the testers knew, had disc front brakes. This, then, must have been an old-stock model, and it seems likely that RMCNA were holding back supplies of the disc-braked cars until they had sold all the drum-braked models; sales were clearly not very brisk if an old-stock car remained unsold some nine months after it had been superseded!

Nevertheless, the testers liked the car, praising its build quality and finish, and admiring the silent progress which allowed the ticking of the clock to be heard below 45 or 50mph. The brakes proved perfectly adequate, and although the steering was a little heavy at low speeds, it was fine on the move. The testers pointed out that there was insufficient room for the middle passenger on the front bench seat, and once again raised the issue of price: 'A delivered price of over $5,000 makes the 3 Liter Rover an expensive car to buy. However, purchase of a Rover is the nearest an individual can come to buying a friend.'

Car Life had certainly hit the nail on the head with its comment about power-assisted steering. As soon as power-assisted steering became optional for the 1961 models, the Engineering Department started looking into the feasibility of retro-fitting 3-litres which had already been shipped to the USA. Americans expected power-assisted steering on cars with luxury pretensions by this stage, and the cars were proving hard to sell without it. Yet the problem of unsaleable cars was only minor compared to another one reported to the Rover Board in mid-November 1960. RMCNA was then losing around £5,000 a week – a substantial sum of money for the time and unfortunately only too indicative of the problems Rover faced across the Atlantic.

THE 3-LITRE STATION WAGON

In autumn 1959, the Rootes Group upgraded its Humber Super Snipe models by enlarging the bores of their 2.6-litre engines to give three litres and 129bhp. This must have worried Rover considerably; the earlier Super Snipe had been a rival for the P4 range, but it was now quite clear that Rootes were challenging the P5 – and a Super Snipe saloon was several hundred pounds cheaper than a Rover 3-litre. Worse, the Super Snipe range also included a capacious and highly-regarded estate car.

Rover therefore asked Pressed Steel to convert a 3-litre into an estate car – or perhaps the idea came from Pressed Steel, who registered the prototype as 432 GWL towards the end of 1959. Pressed Steel's own engineering report on the car was prepared late in 1959 and Solihull had it for examination during the first half of 1960.

The 3-litre Station Wagon (as it is called in the Rover despatch records) was no great beauty, but it was a very practical machine. The fuel tank and rear bulkhead had been removed to give a flat load floor, and the rear bench seat was hinged to fold forwards. A third, rearward-facing, bench seat folded away into the rear floor, and to give adequate headroom for its occupants, the rear roofline was raised above the roof of the main passenger cabin. This terminated in a rather clumsily-styled peak. The rear tailgate hinged downwards just above bumper height (and the overriders were removed to allow this), and the tailgate glass could be wound down into the tailgate by a handle concealed underneath the number-plate plinth. This latter

The 3-litre estate car prototype by Pressed Steel initially had a built-in roof rack and a rather sharp step in the roofline. It also had the three-way colour split associated with the very earliest cars built on the pilot assembly line before the public launch.

Acres of room! The rear seat folded forwards to extend the rear load floor. The hinged floor panel between the two panniers concealing the fuel tanks lifted up to make a rearward-facing bench seat which turned the car into an eight-seater.

arrangement was typical of American estate cars at the time. Twin fuel tanks were fitted in the rear body sides alongside the load area, and they must have been interlinked because there was only one filler cap, behind a flap on the nearside rear wing. Among the other interesting features of the car were that the wheel-changing tools and battery had been relocated under the rear seat, and that a rear compartment heater was also concealed there – anticipating the arrangement on Mk III 3-litres by around five years.

The car was finished in green and white, with the green on the lower body sides also covering the roof to give a three-way colour split. Such colour splits had been seen on the very early production 3-litres, and it could be that 432 GWL was actually converted from one of these early cars, even though it was given its own special commission number of 6309-70001. Former Rover stylist Tony Poole believes that Pressed Steel may have built two P5 estates, but that the company kept the second one and that it was never seen at Solihull.

One way or another, the 3-litre Station Wagon was not a great success. Its styling was not much liked at Solihull, and by July 1960 the roof rack which disguised the step in the roofline had been removed, and the front of the raised rear roof had been smoothed over with clay to effect an improvement in the appearance. The modification was not a success: the car had looked better with its integral roof rack.

Dan Clayton, P5 Project Engineer at the time, remembers the Station Wagon very well,

as I frequently used it at weekends to take my young family to the coast. It was never adopted as a project for production, but was rather more an indulgent 'let's-see' exercise. It was quite spacious in the rear area, but the wind-down tailgate window

was of no use when travelling as it admitted exhaust fumes.

Test engineers Philip Wilson and Brian Terry both remember other problems. As Wilson explained in 1981:

It was too heavy and had a twin tank fuel system which was useless and gave a lot of trouble both in filling and functioning (air locks). The fuel consumption was lousy. Production would have been very costly, and this was possibly the reason for not proceeding further than the prototype.

Terry remembers testing the car with some caution,

because in fact it wasn't our car. It belonged to whoever put the body on. It was converted from an existing saloon, but it did have fairly bad shake. I think that was the reason it wasn't continued. Apart from the fact that, if I remember rightly, it was going to cost a lot more money, about £3,000. The [tailgate] used to wag like fury, and in fact if you went on a fairly rough-surfaced road it would pop open, it was that bad. And the whole thing shook at the back.

For comparison purposes, Rover had a Humber Super Snipe estate on the test fleet. This did not suffer from shake, probably because it was more robustly constructed: 'the Humber weighed a hell of a lot more than the Rover did,' says Terry.

It cannot have taken Rover long to decide against expanding the P5 range to include a Station Wagon variant. 432 GWL nevertheless remained at Solihull until 25 July 1961, when it passed through the Rover Despatch Department on its way back to Pressed Steel at Cowley. No subsequent trace of it – or of the second car, if it existed – has ever been found.

3   The Mk IA Models, 1961–1962

After the 3-litre had been in production for a couple of years, it must have been obvious to Rover that the car was still not right. The lessons learned from the Kenyan proving trips and the Chobham tests in the early part of 1960 had further reinforced the need for further development, and by mid-1960, if not earlier, the work which would produce the very much improved Mk II models in 1962 had begun. In the meantime, however, Rover knew that an interim improved model would be needed if sales were not to continue their 1961-season slide.

So a package of improvements was brought together for announcement in September 1961 as the 1962-season 3-litre. As it happened, Rover was planning to introduce a new car and major unit numbering system at the same time, in which the year of manufacture identifying number would be replaced by a suffix letter identifying design changes of importance when servicing the vehicle or unit. The major changes introduced for the 1962-season products were identified by the suffix letter 'a', and so that year's 3-litres were quite naturally known as 'Mk IA' models. The capital letters were reserved for the model identifications, however: lower-case letters were always used for the component and vehicle number

The 1962-model Mk IA was readily distinguishable from earlier 3-litres. Clearly visible here are the new stainless-steel wheel trims and the quarter-light in the front door. The Lucas 'tripod' headlamps and the front and rear badging were unchanged, however. This car was probably 725-00004a, an early overdrive model for the home market, although it was recorded as 728-00004a (a left-hand drive export model).

Rover 3-litre Mk IA (1961–1962)

Layout
Monocoque bodyshell with front subframe bolted in place. Five- or six-seater saloon, with front engine and rear wheel drive.

Engine
Type	3L7
Block material	Cast iron
Head material	Aluminium alloy
Cylinders	Six, in line
Cooling	Water
Bore and stroke	77.8mm × 105mm
Capacity	2,995cc
Main bearings	Seven
Valves	Inlet valves in cylinder head and exhaust valves in cylinder block
Compression ratio	8.75:1 (7.5:1 for some export markets)
Carburettor	Single SU type HD6 (2in)
Max. power	115bhp gross (105bhp net) at 4,500rpm
Max. torque	164lb.ft at 1,500rpm

Transmission
Manual models	Hydraulically operated single dry plate clutch, 10in diameter
Automatic models	Torque converter

Internal gearbox ratios
Option 1 Four-speed manual

(Final drive 3.9:1)

Top	1.00:1
Third	1.37:1
Second	2.04:1
First	3.37:1
Reverse	2.96:1

Option 2 Four-speed manual with overdrive

(Final drive 4.3:1)

Overdrive	0.77:1
Top	1.00:1
Third	1.37:1
Second	2.04:1
First	3.37:1
Reverse	2.96:1

Option 3 Three-speed automatic (Borg Warner type DG)

(Final drive 3.9:1)

Top	1.00:1
Intermediate	1.43:1
First	2.30:1
Reverse	2.00:1

Suspension and steering	
Front	Independent, with wishbones, laminated torsion bar springs and anti-roll bar
Rear	Semi-floating axle with progressive-rate semi-elliptic leaf springs
Steering	Burman recirculating ball type with variable ratio; power assistance optional
Tyres	6.70 × 15 crossply
Wheels	Five-stud disc type
Rim width	5in

Brakes	
Type	Discs at the front and drums at the rear, with servo assistance
Size	Disc diameter 10.75in Drum diameter 11in

Dimensions (in/mm)	
Track, front	55 (1,397)
Track, rear	56 (1,422)
Wheelbase	110.5 (2,807)
Overall length	186.5 (4,737)
Overall width	70 (1,778)
Overall height	60.25 (1,530)
Unladen weight	3,556lb (1,613kg) (four-speed) 3,612lb (1,638kg) (overdrive) 3,640lb (1,651kg) (automatic)

suffixes, and the same applied to Land Rovers, which took on an a-suffix for 1962 but were known as Series IIA models. 'Please note,' said the *Rover Car Service Newsletter* which introduced the new 3-litres to dealers in September 1961, 'that earlier Rover 3 litre models will be known as the Rover 3 litre Mk I.'

Exterior changes made the M IA models readily recognizable, even at a distance. For a start, there were quarter-vents in the front doors instead of louvres over the front and rear drop-glasses – a change introduced to give draught-free ventilation and respond to customer criticisms that the louvres distorted the view out of the car and were noisy at speed. The second visual change was to the wheel trims, where the P4-style plated hubcaps had given way to

stainless-steel wheel trims which incorporated a slotted outer ring. That 1961 *Service Newsletter* explained that this change was 'to improve appearance', and explained that the new 'hub cover plates' could be fitted retrospectively to later Mk I models. Their design was in fact very similar to that seen on the quarter-scale P5 models some three years earlier, although a turned finish replaced the Mercedes-style painted centre rings which David Bache had proposed.

Less immediately visible were a number of detail changes to the bodyshell, although one of them could be seen from close to. This was the redesigned bumper mountings, which were now slim rods instead of squat brackets, 'to improve appearance and increase strength,' according to that 1961 *Service Newsletter*. It was also necessary to stand very close to a Mk IA to see one of its other modifications – adjustable ball-type washer jets on the chromed air intake panel ahead of the windscreen instead of the fixed jets of the Mk I. As Rover explained, however, the adjustable jets could be fitted to earlier cars if owners so desired.

Many of the Mk IA improvements were electrical, and the whole wiring harness had actually been redesigned to simplify its layout. The cumbersome cable-operated intermediate gear hold of earlier automatic models was replaced on the Mk IA by an electrically-operated gear hold, triggered by a convenient stalk on the steering column. Twin electric fuel pumps replaced the single pump of the Mk I, the second pump both acting as a back-up to the first and offering a reserve capacity. Under the dash, the handbrake warning light introduced on 1961-model automatic 3-litres was added to manual-transmission cars as well, and on all models it was now linked to a float switch in the brake fluid reservoir and doubled as a low-fluid warning light. Last, but by no means least for smokers, the cigar lighter socket was now illuminated so that the element could more easily be replaced at night.

There were just two more changes to the Mk IA. The first was a modified heater with improved air distribution arrangements and water pipes with a larger internal diameter. This change also affected the water-pump outlet, which had a larger internal diameter to suit. The second was that an AC paper-element air cleaner replaced the oil-bath type fitted to Mk I models, probably mainly to simplify routine maintenance.

THE MK IAS ON SALE AND ON TEST

The Mk IA 3-litres were introduced at the Earls Court Motor Show in October 1961,

---

Optional extras available for the Mk IA 3-litre models

Badge bar
Electric immersion heater for cylinder block
Exhaust tail pipe finisher
Extension speaker for radio (fitted under rear parcels shelf)
Floor mats, rubber link type
Fog lamp
Individual front seats, fixed
Individual front seats, 'Lyback' fully adjustable type (from November 1961 approximately)
Laminated windscreen
Pillar pulls
Power-assisted steering
Radio (Pye or Radiomobile)
Removable division by Radford (with bench front seat only)
Roof rack
Seat belts (Irvin) for front and rear
Spot light
Towbar
Two-tone paint
Wing mirrors

---

and publicity material referred to them as the New Series 3-litre Rovers. They were well received by public and press alike, to the great relief of the Rover Board of Directors who were told of this reaction at the board meeting of 14 December 1961. Free service costs were also reported to have been substantially reduced, and the Board saw this as an indication of the improved quality of the car. It was perhaps not before time – although there were of course several further substantial improvements already in the pipeline for introduction on the 1963 models, and in fact the Mk IA 3-litres remained on sale for just twelve months. Rover sold 5,663 of them in that period – plus a handful of smaller-engined P5 variants designed for export only – which represented a massive

Paint and trim colours, Rover 3-litre Mk IA

Note: 2.4-litre and 2.6-litre models were almost certainly offered with the same range of paint and trim options as the 3-litre.

1962 model year

Between its introduction in September 1961 and February 1962, the Mk IA 3-litre was available with the same exterior paint colours and interior trims as the 1961-model Mk I.

Three new paint colours were introduced in March 1962, replacing three existing colours and resulting in a new range of two-tone combinations. The final Mk IA 3-litres were available in ten single-tone colours and in ten two-tone combinations. There were five upholstery colours, in each case with Charcoal Grey trimmings and carpets to match. Wheels were painted Black. The standard combinations for single-tone cars were:

Body colour	Upholstery
Black	Blue, Green, Grey, Red or Tan
Burgundy	Grey or Red
Ivory	Blue, Green, Red or Tan
Light Navy	Blue, Grey or Tan
Medium Grey	Blue, Green, Grey, Red or Tan
Pine Green	Green, Grey or Tan
Shadow Green	Green, Grey or Tan
Slate Grey	Blue, Grey or Red
Smoke Grey	Blue, Green, Grey, Red or Tan
Storm Grey	Blue, Green, Grey, Red or Tan

The combinations for two-tone cars were as follows:

Lower body	Upper body	Upholstery
Black	Smoke Grey	Blue, Green, Grey, Red or Tan
Medium Grey	Burgundy	Grey or Red
Medium Grey	Ivory	Blue, Green, Red or Tan
Medium Grey	Light Navy	Blue, Grey or Tan
Medium Grey	Pine Green	Green, Grey or Tan
Pine Green	Shadow Green	Green, Grey or Tan
Slate Grey	Light Navy	Blue, Grey or Tan
Slate Grey	Storm Grey	Blue, Grey or Red
Storm Grey	Ivory	Blue, Green, Red or Tan
Storm Grey	Medium Grey	Blue, Green, Grey, Red or Tan

30 per cent improvement over the 4,344 3-litres sold during the 1961 season.

The first magazine to publish a road test of the Mk IA 3-litre was *The Motor,* which tested the overdrive-equipped 5124 WD (vehicle number 725-00049a) in its issue dated 22 November 1961.

This improved New Series model offers a high degree of silence and comfort, and above all a combination of good engineering, high-grade finish and careful attention to detail,

Whitewall tyres were fashionable around the turn of the sixties, and Rover showed them on this advertisement for the Mk IA models in the 1961 London Motor Show catalogue. Probably relatively few cars were supplied with them, however; Rover customers tended to be rather conservative!

THE NEW SERIES 3-LITRE ROVER FOR 1962 IS ON STAND NO. 110

Also the Rover '80' and the Rover '100'

was the way the magazine's testers summarized the car. The *Autocar* road test-staff were also very impressed with the revised 3-litre, and said so in their 1 December 1961 test of 4356 WD (car number 730-00059a), an automatic model. 'With both this car and the overdrive model taken to the Continent we have amassed some 3,500 miles in the latest version of Rover's 3-litre. The many staff members who drove either or both of them all agreed that this is quite the best model yet produced by the Solihull factory, which is praise indeed for a company whose vehicles have always set a high standard. The outstanding characteristics of the 3-litre are its comfort and spaciousness, and its silence of running almost regardless of speed.'

Autosport's John Bolster took an overdrive model to the Geneva Motor Show in March

This is the engine bay of a Mk IA model, and shows the black-finished paper-element air cleaner with its twin trumpet intakes. The position of the brake fluid reservoir makes clear that this is a left-hand drive model – in fact, one of the rare Austrian-market 2.4-litre cars. The standard 3-litre engine looked identical from this angle.

1962 and reported on his trip in the maga-zine's issue dated 27 April. He wrote,

> I was a great admirer of the Rover 90 and 105S models, Yet when the 3-litre first came on the scene, I must own to a feeling of dis-appointment. The steering and roadholding were not up to previous standards and the engine was neither so smooth nor so power-ful as I had expected. ... Now, I have really thrashed the latest version across England, France and Switzerland, and I can say that the old quality has been fully regained.

Bolster went on to wax lyrical about the car.

> The whole quality of this car reminds one of the ultra-expensive luxury limousines. The way the doors shut, the standard of the interior trim and upholstery – all these things add up to the pleasure of handling a pedigree car. ... The really fastidious owner will enthuse over the luxury and finish, and he would find it hard to tolerate other less perfect carriages even if their performance were greater ... the Rover 3-litre gives the same pleasure as some prestige cars of far greater cost.

Cars Illustrated for April 1962 tested 4356 WD, the automatic press demonstrator, and that magazine's Douglas Armstrong echoed the enthusiasm shown in earlier tests.

> Like great touring cars of the past, the Rover includes in its specifications every contemporary aid to comfort and fatigue-free motoring, and the quality of its finish and furnishings is quite exceptional in this age of 'simulation'.

Armstrong concluded that,

> With finish, equipment, and engineering to such very high standards the Rover 3-litre is indeed a British car to be proud of. Although a luxury car in every sense of the word it has performance and controllabili-ty in unusually high measure, and must be assessed as great value for money.

However, not every press report was as enthusiastic as *Cars Illustrated* about the performance of the Mk IA's. *The Motor*, find-ing that the maximum speed of their over-drive car was around 97mph, described its performance as only 'satisfactory.' And *Autocar* commented that the automatic ver-sion gave 'reasonably rapid acceleration to 70mph and through to 80mph, but the car then tends to hang fire slightly.' *The Motor* staff also disliked the vagueness of the cranked gear lever, complaining that it was too easy to select reverse instead of first. All these were valid criticisms which Rover, for-tunately, was already planning to address with the Mk II models which would appear in the autumn of 1962.

RIVALS

On the home market, the Mk IA 3-litre's direct rivals were most obviously the big Humbers and BMC's Vanden Plas Princess, although both were actually rather cheaper than the Rover. At the October 1961 Motor Show, the 3-litre with manual transmission was priced at £1,288 and the automatic model at £1,335, in each case plus purchase tax; Humber's Super Snipe Saloon was £1,050, the Super Snipe Limousine £1,150 and the Princess £1,114. Most customers would have recognized that the Rover's handling was far better than the Humber's, although the performance of the two cars was generally similar and their fit and fin-ish were comparable. The Princess was also nothing special in the handling department, although its acceleration was similar to that

Mk IA 3-litre: commission numbers and production figures			
1962 Mk IA	725-00001 to -01612	Manual, home market	(1,612)
	726-00001 to -00190	Manual, export RHD	(190)
	727-00001 to -00306	Manual, CKD RHD	(306)
	728-00001 to -00399	Manual, export LHD	(399)
	730-00001 to -02245	Auto, home market	(2,245)
	731-00001 to -00382	Auto, export RHD	(382)
	733-00001 to -00529	Auto, export LHD	(529)
		Total: 5,663	

of the Rover and the Humber; its biggest drawback was that it was too obviously a badge-engineered Austin A110.

While these cheaper 3-litre luxury saloons undoubtedly took some sales away from the Rover, the Solihull car's pricing just below the 3.8-litre Mk 2 Jaguar (at £1,360 with automatic transmission) must also have earned it a few sales from those who appreciated fine engineering and creature comforts more than the very much greater performance which the Jaguar offered. Also priced around the same in Britain were the Mercedes-Benz 190 (at £1,362) and the Auto Union 1000SP Coupé, although neither of these German cars was strictly in the same class as the Rover and it was only heavy import taxes which put them into the same price bracket.

The picture was of course different in overseas markets, where the Rover was always a high-priced import and was never able to compete on equal terms with home-grown luxury models.

PRODUCTION CHANGES

The twelve months of the Mk IA models' production saw Rover make the usual stream of minor specification changes, and full details are given in the table. Noise reduction and the braking system were two of the main focuses of the engineers' attention. Particularly interesting, however, was the introduction of safety belt kits for dealer fitment in December 1961 and their availability as a line-installed option from the following July. At this stage, safety had become a preoccupation at Solihull, and the forthcoming P6 saloon was being designed with crash safety firmly in mind. This was of course several years before Ralph Nader's crusades alerted US legislators to the need for safety legislation governing cars, and in fact only Mercedes-Benz and Volvo had taken crash safety at all seriously by this stage. Rover was therefore something of a pioneer in this area, although whether the company started looking at safety simply to keep up with the German and Swedish cars, which were its rivals in some markets, is not clear.

The reclining front seats introduced as an option in November 1961 made a worthwhile improvement to the 3-litre's specification, and no doubt Rover would have announced them at the beginning of the season in September if they had been ready in time. Strangely, they seem not to have been mentioned in sales literature for the car until March 1962, at which time three new paint colours – Burgundy, Light Navy and Pine Green – also brightened up the cars' appearance. The other seven exterior colours and all the interior trim colours nevertheless remained unchanged

Rover 3-litre Mk IA: production changes

Note: All dates given for changes where no chassis number is quoted must be treated as approximate. The dates are taken from issues of *Rover Service Newsletter*, which usually reported changes between one and four months after they had actually been made on the assembly lines. Dates given for changes where a chassis number is quoted are exact for home market models and reflect the date into despatch (i.e. the date the car was transferred from the assembly lines to the Despatch Department) given in records held by the British Motor Industry Heritage Trust.

Date	Home, manual	Home, auto	Other	Remarks
September 1961	Mk IA models introduced			
	725-00085a	730-00125a	726-00023a 728-00013a 731-00021a 733-00024a	Front brake callipers modified, with improved seals.
October 1961	725-00323a	730-00432a	726-00046a 728-00025a 731-00051a 733-00064a	Modified air cleaner and elbow, to prevent rattle.
November 1961	725-00341a	730-00501a	726-00055a 728-00030a 731-00061a 733-00086a	Zone-toughened windscreen fitted.
	725-00343a	730-00514a	726-00056a 728-00032a 731-00062a 733-00092a	5½in diameter rear exhaust silencer introduced, to prevent boom.
		730-00545a	731-00063a 733-00107a	Auto selector rod with ball ends replaced rod with clips, to prevent buzz.
	725-00393a	730-00573a	726-00061a 728-00041a 731-00077a 733-00117a	Improved dust sealing of bodies.
				Rectified clock fitted (Smith's CE.3110/02).
				Reclining front seat option introduced.
December 1961				Front and rear safety harness kits available.
January 1962				Modified fixing for lower gear selection lever on automatic cars.
February 1962	725-00722a		726-00113a 728-00137a	Strengthened gear lever with increased radius above the spherical seating.

				Improved Girling brake servo, identifiable by eight-bolt end cover fixing (older Mk II has seven bolts) and X stamped on body; retro-fit possible.
March 1962	725-00803a		726-00698a	Synthetic rubber oil seal for rear hub bearing replaced leather seal.
	725-00805a	730-01158a	726-00117a 728-00157a 731-00219a 733-00257a	New front exhaust pipe mounting with packing between bracket and plate to improve flexibility of down pipe mounting and prevent stress on manifold; retro-fit possible.
	725-00816a	730-01173a	726-00119a 728-00156a 731-00220a 733-00254a	Improved steering relay, with Ferodo friction discs in place of wooden discs; retro-fit possible.
	725-00859a	730-01246a	726-00124a 728-00175a 731-00231a 733-00290a	Don 105/3 brake pad material replaced Ferodo DS5S, to prevent pads glazing and sticking when wet; retro-fit possible.
				Door skins made available as service replacements.
				Towbar introduced as option for Mk IA; new type suits Mk I as well.
				New paint colours: Pine Green (replaced Rush Green), Light Navy (replaced Royal Blue) and Burgundy (replaced Norse Blue).
April 1962	2.6-litre models introduced (date approximate)			
				Improved crankshaft oil seal with new seal retainer halves and separate split oil seal and garter spring, from engines 725-00021a, 726-00771a, 727-00033a, 728-00256a, 730-00998a and 731-00233a; retro-fit possible.
				Larger radius on ball ends of pushrods to minimize wear; inlet cam followers modified to suit; modified exhaust valve roller followers, from engines with suffix 'b'.

		Third fixing added to brake disc shields to improve seal between shields and stub axles; retro-fit possible.
May 1962		Anti-rattle springs introduced as service fitment for models with disc brakes.
		Insulator added to clutch slave cylinder piston rod, replacing earlier clevis jaw, to prevent transmission of engine noise into body through clutch linkage; retro-fit possible.
		Modified steering column with bearings at top and bottom of tube, to prevent steering column rattle and simplify assembly; new column has a rubber grommet at its base inside the car.
		Starter switch with 'gated' action introduced (key has to be pushed in to operate starter); retro-fit possible.
July 1962	2.4-litre models introduced (date approximate)	
		Front seat safety harnesses introduced as an option.
		New exhaust manifold gaskets, with material facing on one side only; retro-fit possible.
		Champion UN-12Y sparking plugs recommended as alternatives to standard N5 plugs when fouling occurs.
		Modified steering box with spring-loaded control for rocker-shaft end-float in place of screw adjuster; retro-fit possible.
August 1962		Modified mainshaft lubrication valve retaining ring on Borg Warner transmission; modified units had 'R' stamped on rear end of mainshaft.
		Additional copper washer on set bolt fixing brake cylinder and plates to main gearbox casing on

	Borg Warner transmission, to prevent oil seepage (from transmission no.10024).
	Cibie 'Diplomat' replacement headlight units available, allowing instant switch from left-hand to right-hand dip.
	Electrical items now supplied separately from towbar kits.
September 1962	Window winder handles with black plastic grips replaced winders with metal grips.

The interiors of the last Mk I and the Mk IA models had lost the contrasting charcoal grey panels. Compare this Mk IA door trim with the Mk I type illustrated on page 34.

from the 1961 season. The final cosmetic change for the Mk IA occurred right at the very end of production, when some cars were fitted with the black plastic window-winder grips destined for the Mk II models instead of with the standard metal grips. It is not clear how many cars were affected, but there were probably very few.

THE MK IA IN THE USA

The Mk IA cars did not sell significantly better than their Mk I forebears in the USA, and fared much the same in the hands of road testers. *Motor Trend* tested an automatic model for its June 1962 issue, and singled out the car's build quality, brakes and ride for particular praise. The engine was 'one of the smoothest, most flexible that we've come across', but performance was only average with 60mph coming up from rest only after 20 seconds. The magazine's testers also liked the instrument layout, but thought the car should have an oil pressure gauge rather than a simple 'idiot light' to warn of problems. The basic price of a 3-litre was still $4,995, but the 'as-tested' price, with power-assisted steering, automatic transmission, whitewall tyres and a radio was a massive $5,495. With a well-equipped base-model 325bhp Cadillac 62 hardtop sedan costing just over $5,250, it is small wonder that the 3-litre appealed to relatively few American customers.

THE CKD 3-LITRES

Ever since the early fifties, Rover had served a number of overseas markets with cars or Land Rovers which were manufactured in Britain but were shipped abroad as CKD (Completely Knocked Down) kits of parts for assembly in the country of destination. This was a way of avoiding punitive import taxes on fully-assembled vehicles, and was welcomed by overseas governments because it provided jobs for the local workforce. In some cases, where tyres, glass, paint, electrical parts and other components were sourced locally, it also supported local industry.

No 3-litres were shipped out from Solihull in CKD form during the period of the Mk I models' production. This may well have been because the early cars gave so much trouble in service and Solihull considered it wiser not to introduce the additional risks associated with overseas assembly. Nevertheless, by the time of the much-improved Mk IA models, the company was prepared to embark on CKD operations for the P5.

Just 306 Mk IA 3-litres were shipped abroad as CKD packs, and all of them went to Motor Packing in South Africa between 19 September 1961 and 29 June 1962. No detailed information is available about the specification of these cars.

THE 2.6-LITRE P5

As Chapter 1 explained, Maurice Wilks had always envisaged the P5 not as a single model but as a whole range of models. Even though he had agreed that the car should

enter production with just one size of engine, he soon initiated a project for a second body style – and he was almost certainly behind the discussion about broadening the range which took place on 19 September 1958 at an Engineering Board meeting about future developments of P5. He had simply bided his time, waiting until the P5 had entered production before raising the issue again!

What the Engineering Board considered in September 1958 was an economy model P5, to be introduced during 1959 with a 2-litre four-cylinder engine. As the timescale for development was short, no doubt the engine they had in mind was the 60bhp 1,997cc engine of the P4 60, as the 90bhp overhead-camshaft 1978cc engine destined for the P6 was still in the prototype stage. However, the idea of a 2-litre P5 was short-lived, and by the time the Engineering Board met on 1 January 1959, the economy model was envisaged as a six-cylinder car with the new 2,625cc seven-bearing engine. This was simply a short-stroke edition of the 3-litre engine, and was planned for introduction in the P4 100 later that year.

No doubt the pressure of other projects ensured that this one was put on the back-burner for a time; Solihull was heavily preoccupied with the P6 project at the beginning of the sixties, and there were Land Rover developments in hand as well as the building of a new manufacturing plant at Pengam in Wales and the threat that Land Rover production might actually be moved down there to make way for P6 production at Solihull. However, when the opportunity arose to look again at a 2.6-litre P5, the Rover engineers did so. The catalyst for their renewed interest was the possibility of increasing P5 sales in France.

French taxation policy in the early sixties meant that cars with engines bigger than about 2.7 litres were prohibitively expensive to run, and this inevitably restricted sales of

---

Rover 2.6-litre Mk IA Saloon (1962)

Specification as for contemporary 3-litre with manual transmission, except:

Engine

Type	Short-stroke 3L7
Block material	Cast iron
Head material	Aluminium alloy
Cylinders	Six, in line
Cooling	Water
Bore and stroke	77.8mm × 92.075mm
Capacity	2,625cc
Main bearings	Seven
Valves	Inlet valves in cylinder head and exhaust valves in cylinder block
Compression ratio	8.8:1 (7.8:1 optional)
Carburettor	Single SU, type not known
Max. power	123bhp at 5,000rpm (121bhp at 5,000rpm with 7.8:1 compression)
Max. torque	142lb.ft at 3,000rpm (136lb.ft at 3,000rpm with 7.8:1 compression)

Transmission

All models had manual transmission, but it is not clear whether overdrive was standard or not.

---

the 3-litre in that country. The 2,625cc six-cylinder engine, however, fitted neatly below that tax barrier. Whether the idea came from Rover or from its French importer, Franco-Britannic Autos of Paris, by the end of 1961 a plan had been drawn up to build a 2.6-litre version of the P5 for the French market.

In its original production form for the 1960-model P4 100, the 2,625cc engine produced just 104bhp. However, under development during 1961 was a 123bhp version, using the Weslake cylinder head originally designed for the 3-litre and due to be announced in the autumn of 1962. It was this engine – which later appeared in the P4 110 – which Rover decided to use for the French car. With 123bhp, it was actually more powerful than the existing 115bhp production 3-litre engine, although its torque of 142lb.ft at 3,000rpm could not match the 3-litre's 164lb.ft at 1,500rpm.

There was little point in building proto-types, as the 2.6-litre Weslake-head engine had already been developed to production

---

Mk IA 2.6-litre: commission numbers and production figures

1962 Mk IA	756-00001 (one only)	Manual, export RHD	(1)
2.6-litre	758-00001 to -00024	Manual, export LHD	(24)
			Total: 25

readiness and the P5 into which it would be put needed no modification from the existing production type. So Rover built its first 2.6-litre P5 at the end of January 1962. This was a blue car with right-hand drive, and was retained at Solihull for testing. Test engineer Brian Terry remembers that it earned itself a reputation for having an unburstable engine, which boded well for the production examples. It was also the very first production Rover of any kind to have a Weslake-head engine: as far as most markets were concerned, the Weslake-head engines did not exist until September 1962, when they were seen on the P4 110 and the Mk II 3-litres.

However, no more right-hand drive 2.6-litre Mk IA models were built. Production of left-hand drive cars started at the end of April and twenty-five were built between then and the middle of April. All went to Franco-Britannic Autos in Paris, who found enough enthusiastic customers for Rover to agree to build Mk II versions for the 1963 season.

These first 2.6-litre P5s were interesting hybrids, being essentially left-hand drive Mk IAs with certain Mk II features. The Weslake cylinder head was of course a Mk II feature, and the grey-enamelled grille badge also anticipated the badge used on the Mk II. This badge of course described the car as a 'Rover 2.6-litre'.

THE 2.4-LITRE P5

Probably encouraged by the ease with which the 2.6-litre P5 Project had come together, Rover was receptive when it was asked to build another small-engined P5. However, this one was very much less successful. Just twenty-five examples were built, and the special engine never appeared in another Solihull product and therefore proved very expensive to build.

In the early sixties, Rover's Austrian importer was a company called O.J. Aulehla,

---

Rover 2.4-litre Mk IA Saloon (1962)

Specification as for contemporary 3-litre with manual transmission, except:

Engine
Type	Short-stroke 3L7
Block material	Cast iron
Head material	Aluminium alloy
Cylinders	Six, in line
Cooling	Water
Bore and stroke	77.8mm × 87.3mm
Capacity	2,490cc
Main bearings	Seven
Valves	Inlet valves in cylinder head and exhaust valves in cylinder block
Compression ratio	8.8:1
Carburettor	Single SU, type not known
Max. power	approximately 109bhp
Max. torque	Not known

Transmission
All models had manual transmission, but it is not clear whether overdrive was standard or not.

with headquarters in Vienna. Tax regulations in Austria penalized large-engined cars, but it appears that the head of the company became convinced that it would be possible to sell the P5 in worthwhile quantities in Austria if only it had an engine capacity below 2.5 litres.

Some time in 1961, he managed to convince Rover's Production Director, William Martin-Hurst, that such a car had a future, and engine designer Jack Swaine remembers being asked by Martin-Hurst to develop a suitable engine from the existing six-cylinder block. The simple solution was to provide a new short-stroke crankshaft, and Swaine's team came up with an engine which displaced 2,490cc. Like the 2.6-litre engine in the French P5s, it had the Weslake-developed cylinder head to maximize power and torque. Factory figures are not available, although a registration document associated with one of the surviving cars gives the power output as 111PS (roughly 109bhp) at 5,000rpm. This engine was put into a model called the Rover 2.4-litre, which entered production in July 1962, a week before assembly of the last 2.6-litre P5s finished. Rover built just twenty-five cars – exactly as it had for the 2.6-litre – and assembly finished at the end of August. Like the 2.6-litre model, the 2.4-litre was essentially a Mk IA but anticipated the Mk II specification with items like its grey-enamelled grille badge (which of course read 'Rover 2.4-litre'). The chassis number plate on the door pillar further identified the car as a Rover 2.4-litre Mk IA.

All twenty-five cars were delivered to O.J. Aulehla, but they appear not to have been the

Just twenty-five Mk IA 2.4-litre models were built, and the only external means of recognition was this special grille badge. The 3-litre cars still had the Mk I type of cast grille badge, and they would not take on this style of badge until Mk II production began.

success the importer expected. Of the three survivors known today, one was not registered until March 1963 – so it waited some six months to find a buyer. A second is Herr Aulehla's own car. This latter has power-assisted steering, although it appears that not all the 2.4-litre models were so equipped. All the cars did have manual gearboxes, however, probably with overdrive as standard. By 1964, O.J. Aulehla had been replaced as the company's Austrian importer by Carl Jeschek of Vienna, and it is not clear whether a Mk II version of the car ever went into production, although such a model was certainly planned, as Chapter 4 reveals.

Mk IA 2.4-litre: commission numbers and production figures

1962 Mk IA 2.4-litre	788-00001 to -00025	Manual, export LHD	(25)
			Total: 25

4 The Mk II Saloons, 1962–1965

It would probably be true to say that the 3-litre's classic period was between 1962 and 1965, when the Mk II Saloons were in production alongside the new Coupé companion model (which is covered in Chapter 5). With these cars, the shortcomings of the early 3-litres were eliminated, and the competitive models put forward by other manufacturers were unable to make a significant dent in the Rover sales figures. The Mk II 3-litre was a spacious, refined, fast and beautifully built luxury saloon – and some of its qualities were never surpassed by later versions of the P5 range.

The buying public appreciated these new cars as well. A comparison of the sales figures for the Mk II Saloons with those for Mk IA models suggests that the newer cars did not sell as well, but adding the Mk II Saloon figures to those for the Coupé makes clear that the P5 was doing better than ever. The Mk II got off to a moderate start, but sales soon began to pick up as news about the new models' improvements spread. Overall totals for the 1963 season were just under 4 per cent down on totals for 1962, but 1964 saw sales increase by a massive 42 per cent. That year broke the previous record for P5 sales, set in

With its suspension lowered by an inch all round, the Mk II 3-litre Saloon looked much more purposeful than earlier 3-litres. This is one of the earliest press pictures of the car, issued in 1962.

Rover 3-litre Mk II Saloon (1962–1965)

Layout
Monocoque bodyshell with front subframe bolted in place. Five- or six-seater saloon, with front engine and rear wheel drive.

Engine

Type	3L7 with Weslake head
Block material	Cast iron
Head material	Aluminium alloy
Cylinders	Six, in line
Cooling	Water
Bore and stroke	77.8mm × 105mm
Capacity	2,995cc
Main bearings	Seven
Valves	Inlet valves in cylinder head and exhaust valves in cylinder block
Compression ratio	8.75:1 with manual transmission 8:1 with automatic transmission
Carburettor	Single SU type HD6 (2in)
Max. power	134bhp gross (121bhp net) at 5,000rpm with manual transmission 129bhp gross (116bhp net) at 4,750rpm with automatic transmission
Max. torque	169lb.ft at 1,750rpm with manual transmission 161lb.ft at 3,000rpm with automatic transmission

Transmission

Manual models	Hydraulically operated single dry plate clutch, 10in diameter
Automatic models	Torque converter

Internal gearbox ratios
Option 1 Four-speed manual with overdrive

	(Final drive 4.3:1)
Overdrive	0.77:1
Top	1.00:1
Third	1.27:1
Second	1.88:1
First	3.37:1
Reverse	2.96:1

Option 2 Three-speed automatic (Borg Warner type DG)

	(Final drive 3.9:1)
Top	1.00:1
Intermediate	1.43:1
First	2.30:1
Reverse	2.00:1

Suspension and steering

Front	Independent, with wishbones, laminated torsion bar springs and anti-roll bar
Rear	Semi-floating axle with progressive-rate semi-elliptic leaf springs
Steering	Burman recirculating ball type with variable ratio and 17.6:1 ratio at straight ahead position; power assistance optional ('a' suffix cars, to November 1962)
	Burman recirculating ball type with variable ratio and 20.3:1 ratio at straight ahead position; power assistance optional ('b' suffix cars, December 1962 to February 1964)
	Burman recirculating ball type with variable ratio and power assistance ('c' suffix cars, March 1964 on)
Tyres	6.70 × 15 crossply
Wheels	Five-stud disc type
Rim width	5in

Brakes

Type	Discs at the front and drums at the rear, with servo assistance
Size	Disc diameter 10.75in
	Drum diameter 11in

Dimensions (in/mm)

Track, front	55 (1,397)
Track, rear	56 (1,422)
Wheelbase	110.5 (2,807)
Overall length	186.5 (4,737)
Overall width	70 (1,778)
Overall height	59.25 (1,505) (Saloon)
Unladen weight	3,640lb (1,651kg) (overdrive)
	3,654lb (1,657kg) (automatic)

1960, and would remain the P5's best-ever year. So even though overall P5 sales dropped by nearly 14 per cent in the 1965 season, the actual sales totals still remained very healthy.

A NEW ENGINE AND GEARBOX

Rover started looking for more power from the 3-litre engine quite early on, and former development engineer Brian Terry remembers that the main cause of its restricted output was identified as a breathing deficiency. The Experimental Department tried out a variety of solutions, including the multiple-carburettor installations which Managing Director Maurice Wilks and Chief Engine Designer Jack Swaine both mistrusted on the grounds that the tuning problems they caused often outweighed the advantages they brought. There were always other difficulties with these, however, not the least of them being

The air cleaner of the Weslake-head engine always had an aluminium finish instead of the black found on Mk IA models. This is a suffix 'c' Mk II Saloon.

The Mk II models at Earls Court, October 1962

In its issue dated 19 October 1962, *Autocar* magazine printed this evocative description of the Rover stand – number 117 – at the 1962 London Motor Show:

Special plinths display the new Rover 3-litre Coupé and 3-litre Mk II Saloon. The Coupé revolves on a slender support, the base of which is covered with line drawings of Rover cars. Alongside, the 3-litre Saloon is static on its plinth. On the stand floor for close inspection there is another example of the 3-litre Coupé with manual transmission, and an automatic Mk II finished in Burgundy and Stone. Power steering, which is a standard feature of the new Coupé, is also fitted to the saloon exhibited.

Also on the Rover stand were a two-tone green 110 and a Charcoal Grey 95, both models new to the P4 range that autumn.

that the space they would have taken up under the 3-litre's bonnet would have demanded a rearrangement of various ancillaries.

Nevertheless, various single-, twin-, triple- and even six-carburettor installations were tried on the test-beds, using wooden cylinder heads and manifolds lined with a spray of asbestos so that a full-throttle power test could be done before the asbestos cracked and heat burned up the wooden components. One triple-carburettor engine which Brian Terry remembers testing gave more than 160bhp on the test-bed, using a special exhaust and the standard camshaft but different timing. Unfortunately, though, it was less efficient at low speeds than the production single-carburettor engine. Former racing driver Raymond Mays, who had close links with the Rover Company, also designed a triple-carburettor manifold for the 3-litre, and submitted it to Solihull for testing. It would idle beautifully, as low as 400rpm, remembers Brian Terry, and 'the bottom end performance was excellent, but we thought it rather strange that it didn't seem to go any faster. We put it on the test-bed and in fact it turned out 1½bhp more than the single-carburettor which we'd done!'

The Graber Convertible

The small coachbuilders survived on the European continent long after heavy taxation and the difficulties associated with monocoque construction had forced most of their British counterparts out of business. Among them was the respected Swiss firm of Graber, based at Wichtrach near Bern. At the Geneva Motor Show in March 1963, Graber displayed a drophead conversion of the Rover 3-litre alongside a number of special-bodied Alvis 3-litres. Whether the car was built to meet a customer's order or was simply a Motor Show special to demonstrate the coachbuilder's art is not clear. One way or another, it remained unique.

Like the Chapron-converted car commissioned by Rover the previous year (*see* Chapter 2), the Graber car had two doors instead of the four of the saloon from which it was converted. Its convertible top was rather less attractive than the Chapron car's, however, having rather heavy rear quarters in the tradition of the Swiss and German coachbuilders. The side trim strips of the original car were also replaced by a broad rubbing-strip which ran from the front indicators to the rear light clusters, and the car was finished in dark blue. It had been re-upholstered in blue, with different pleating from the Rover style. Graber displayed it at Geneva with wide-band whitewall tyres.

According to a report in *Autocar* for 22 March 1963,

> each door is made much wider to give good access to the rear seats and additional panel and frame-work is added to fill the gaps left by the discarded rear doors; this also restores body stiffness in the absence of a steel roof. Frameless windows are fitted to each door and the hood stows neatly in a recess behind the rear seat.

The magazine also commented that Graber's bodywork, 'particularly the panelling, fit of doors and paintwork, compares well with anything produced elsewhere in the world.'

The Graber car was converted from an early left-hand drive saloon with overdrive, car number 773-000017a. Records held by the British Motor Industry Heritage Trust show that it left the production line at Rover on 22 October 1962 and was shipped out to Fehlmann and Company, Rover's Swiss importers, three days later. As originally built, the car was finished in Stone Grey.

The Graber 3-litre still survives in Switzerland in the hands of a Rover enthusiast, but by 1996 it had been off the road for several years and was awaiting restoration.

So the Raymond Mays manifold never did go into production, but the prototype still survives in modified form in the Rover Special single-seater racing car built in the late forties by a group of Rover engineers.

In the end, the Rover engines team came up with a redesigned induction system in which the inlet manifold and cylinder head were cast as separate units and bolted together. This gave a worthwhile power increase with the simplicity of a single SU carburettor. To make doubly sure, Maurice Wilks called in performance development expert Harry Weslake, to see if he could improve the design even further. The final result gave a 16 per cent power increase over the original 3-litre engine, and the revised engine was always known as the 'Weslake head' type. It gave the extra performance which Rover wanted, but its torque characteristics were not ideally suited to the Borg Warner DG automatic gearbox. So Rover came up with a different camshaft and a lower compression ratio which boosted torque at the expense of top end power, and fitted this to engines destined for Mk II automatic cars.

Meanwhile, to get the best out of the engine with a manual transmission, the

Graber's 3-litre convertible was in a semi-derelict condition when this picture was taken in the late seventies. The thick side rubbing-strip is clearly visible, and the special upholstery with its deep pleats can just be seen.

four-speed gearbox was also modified. Strangely – perhaps because the budget was restricted – Rover did not take the opportunity to add synchromesh to first gear, but the engineers did raise the ratios of second and third gears to allow higher speeds in these ratios. The new close-ratio gearbox was accompanied by a new short-stick remote-control gearchange, which removed all the vagueness associated with the original cranked lever, known to its friends and enemies alike as the 'mustard spoon'. At the same time, the overdrive kick-down facility was deleted to prevent drivers from kicking down into direct top at the higher speeds which the Mk II could attain, and so overspeeding the engine.

THE REST OF THE MK II PACKAGE

This improved performance was matched by changes to the suspension and steering of the 3-litre. The Mk II models rode an inch lower on their suspension than earlier cars, which lowered their centre of gravity and so reduced cornering roll. New tyres – Dunlop RS5 or Avon Turbospeed Mk IV – gave better grip on corners and finally eliminated the tyre squeal associated with the Dunlop Gold Seal tyres fitted to many earlier cars. Higher pressures of 26psi all round instead of the 22–24psi recommended on the Mk I and Mk IA also contributed to the overall improvement. Nylon buttons added between the leaves of the rear springs reduced friction and wear, and modified damper settings were claimed to give more consistent ride control over a longer period. Reduced maintenance was also a feature of the Mk II cars, which needed the attention of a grease gun on only one nipple (on the propshaft centre bearing) every 6,000 miles.

It was obviously important from a sales and marketing point of view that the new models should look different from the old, but Rover's options seem to have been limited by the budget which was available. Stylist David Bache put forward a proposal for a new honeycomb-pattern grille, not unlike the one he had incorporated on the 1955 full-size mock-up 'ROV 575', and planned to use a similar style of grille on the new P4 models (the 95 and 110) which were to be introduced at the same time as the Mk II 3-litres. Rover's grille suppliers, Joseph Fray Ltd, had manufacturing drawings for the new grilles ready by the early part of 1962, but in the end the project did not go ahead. An engineering report from Pressed Steel, prepared in October 1962, also shows the Mk II Saloon as having stainless steel window frames. These never entered production, either, although it is not clear whether they were cancelled for budgetary reasons or someone at Pressed Steel confused the Mk II Saloon with the new Coupé, which of course *did* have stainless-steel door frames.

One way or another, the exterior changes to the production Mk II Saloons ended up being very limited. There were different badges on the grille, where a grey enamelled 'Rover 3-litre' plate replaced the ribbed metal type, and on the bootlid, where individual letters replaced the underlined Rover name

---

Paint and trim colours, Rover 3-litre Mk II

1963–1966 model years

There were ten standard paint colours and six standard interior colours with carpets to match. Ten two-tone paintwork combinations were available. The standard combinations for single-tone cars were:

Body colour	Upholstery
Black	Blue, Green, Red, Stone or Tan
Burgundy	Grey, Red or Stone
Charcoal	Blue, Green, Grey, Red or Stone
Juniper Green	Green, Grey, Stone or Tan
Light Navy	Blue, Grey or Stone
Marine Grey	Blue, Green, Grey, Red, Stone or Tan
Pine Green	Green, Grey, Stone or Tan
Stone Grey	Blue, Green, Red or Tan
Steel Blue	Blue, Grey, Red or Tan
White	Blue, Green, Red, Stone or Tan

The combinations for two-tone Saloons were as follows:

Lower body	Upper body	Upholstery
Charcoal	Marine Grey	Blue, Green, Grey, Red, Stone or Tan
Juniper Green	Pine Green	Green, Grey, Stone or Tan
Light Navy	Steel Blue	Blue, Grey or Stone
Marine Grey	Black	Blue, Green, Grey, Red, Tan or Stone
Marine Grey	Light Navy	Blue, Grey or Stone
Marine Grey	White	Blue, Green, Red, Stone or Tan
Pine Green	Marine Grey	Green, Grey, Stone or Tan
Steel Blue	Charcoal	Blue, Grey or Red
Stone Grey	Burgundy	Grey, Red or Stone
Stone Grey	Juniper Green	Green, Grey, Stone or Tan

(Right) *A neat grey enamelled grille badge characterized the Mk II cars. The triangular Diakon plastic badge on the grille's centre spine was unchanged from the previous type, however.*

(Left) *Badging had changed at the rear, too. The Rover name was no longer underlined but made of separate letters, there was a '3 litre' script badge alongside the number plate box, and automatic models had an additional badge.*

and a new '3 litre' script badge was added. Automatic cars had an additional identifying badge below this. The lowered suspension also made the cars look rather more purposeful than the Mk IA they replaced, but it was quite difficult to tell Mk IA and Mk II 3-litres apart at a distance.

The interior changes were far more radical, however. Although the seats themselves remained unchanged, almost everything around them was different from its Mk IA counterpart. Most noticeable was that David Bache had dispensed with the slightly untidy look of the door trims, and had styled neat new panels with large map pockets at their lower edges. Similarly, the matching charcoal grey plastic of the steering wheel and steering column shroud had given way to black, which toned in with the instrument panel and crash padding. Redesigned

armrests also improved the appearance of the doors, and the window winders had the black plastic grips first seen on the final Mk IA models over the summer of 1962.

On the facia and around the door windows, heavily varnished African cherrywood had replaced the walnut used in earlier 3-litres, and the prominent glovebox handles had been made smaller and incorporated in a neat stainless-steel trim strip which mated up with the finisher in the wooden door cappings. Meanwhile, crushable sun visors, trimmed to match the head lining, replaced the tinted plastic type which must have been seen as a safety hazard. And between the front seats, the new remote-control gear lever left room for a moulded carpet section designed for use as a parcels tray – although it also meant that the bench front seat could no longer be had with manual transmission.

The substantially revised interiors of the Mk II cars brought plainer door trims with different armrests. This Mk II Saloon displays the optional Lyback front seats and shows quite clearly how untidy the original Irvin static seat belts looked, even when stowed on the hooks provided for them on the B-pillars.

The Panelcraft Convertible

A third 3-litre convertible was made from a Mk II Saloon by the London coachbuilders FLM (Panelcraft) over the winter of 1963–1964. It was commissioned by George Hansson, a Swedish civil engineer then living in Britain, and the conversion cost £700 – a little over half the cost of the basic Saloon. Hansson was a friend of Rover's Technical Director, Peter Wilks, and took the car up to Solihull during April 1964 so that Rover could evaluate the work which had gone into it.

Wilks decided against pursuing the idea of a 3-litre convertible, but the following year, when Rover was thinking about a convertible based on the 2000, Panelcraft were commissioned to build an evaluation prototype. No convertible 2000 ever went into production, either, but Panelcraft did build on their relationship with Rover in later years. By the end of the sixties, the company had persuaded Rover to honour its standard warranty on Panelcraft's estate-car conversion of the P6 saloon, and during the seventies Panelcraft was one of the first companies to convert Range Rovers to four-door configuration for the Middle East, this time enjoying the coveted Land Rover Approval.

Like the Chapron and Graber 3-litre convertibles, the Panelcraft car had just two doors, each longer than the standard saloon door. Its unique feature, however, was an additional drop-glass in the body side behind the door, which made the rear seat very much less claustrophobic when the hood was up. The car probably looked its best with all the side windows raised and the top down, the hood then being concealed under a neat cover which stood proud of the bodywork.

The Panelcraft car was converted from a Home Market 3-litre Mk II Saloon with overdrive, number 770-02021b. It came off the production line on 23 October 1963, and was despatched out to Atkinson and Company, a Rover dealer in Huddersfield, on November 1st. Panelcraft's conversion retained the car's original colours of Light Navy with blue seats, although the passenger seat was modified to tip forwards for access to the rear and the rear seat was made narrower to allow for the folding mechanism of the hood.

George Hansson sold the car on in 1967. It disappeared from view in the mid-seventies and was thought to have been scrapped, but it resurfaced in 1994 after nearly 20 years in storage and has now been carefully restored for its enthusiastic new owner.

The 3-litre convertible built for a private customer by FLM Panelcraft was the only one of its kind made in Britain.

The whole instrument panel had also been thoroughly revised, although it looked generally similar to the earlier type. Its front panel was now covered in matt black leathercloth, and matt black bezels around the main dials and the clock replaced the chromed ones of the earlier cars. There were matt black spade-shaped switches instead of the glossy pointed type, and there had been a complicated series of changes to the switchgear. The main lighting switch now incorporated a 'parking lights' position, in which all lights were extinguished except the front sidelight and the tail light on the driver's side. The heater fan switch – now offering two speeds – was added to the instrument panel and the function of the sliding control alongside the centre glove-box (or radio) was altered to suit. The new fan switch displaced the oil level indicator switch, which now moved to the bottom of the instrument panel, where it displaced the headlamp and indicator telltales. These in turn were now located at the bottom of

the speedometer. Away from the instrument panel itself, a headlamp flasher was now incorporated in the indicator stalk mounted on the steering column, and a large rectangular amber brake warning light replaced the small round red one of the earlier cars.

Detail specification revisions completed the picture. The Mk II had much-improved windscreen-wiper arrangements, with longer blades to sweep a larger area of the screen

This three-position switch on the instrument binnacle was fitted only to c-suffix versions of the Mk II 3-litre.

Optional extras available for the Mk II 3-litre models

Badge bar
Dipping interior mirror (from October 1965)
Electric immersion heater for cylinder block
Exhaust tail pipe finisher
Extension speaker for radio (fitted under rear parcels shelf)
Floor mats, beige nylon fur (from October 1965)
Floor mats, charcoal grey rubber (from October 1965)
Floor mats, rubber link type (to September 1965)
Fog lamp
Foot pump
Heated rear window
Individual front seats, fixed (standard on manual-transmission cars)
Individual front seats, Lyback fully adjustable type
Laminated windscreen
Mud flaps, front and rear (from October 1964)
Pillar pulls
Power-assisted steering (standard from October 1964)
Radio (Pye or Radiomobile) (to September 1964)
Radio (Pye Medium/Long Wave, Pye Medium/Short Wave, Pye portable Medium/Long Wave, or Radiomobile Medium/Long Wave) (from October 1964)
Removable division by Radford (for models with bench seat only)
Removable division by Radford (for cars with fixed individual front seats) (from October 1964)
Roof rack
Seat belts (Irvin) for front and rear
Special paint finish or interior trim to customer's requirements
Spot light
Towbar
Two-tone paint
Whitewall tyres
Wing mirrors

and a new motor which incorporated a thermostatic overload cutout. There was a new washer bottle, too, now made of plastic instead of glass. Customers could now order their 3-litres with an electrically-heated demister element in the rear window, an extra-cost option which was still rare even in the luxury car market. More powerful sealed-beam headlamps replaced the 'tripod' type with their separate bulbs, and to cope with all the additional electrical demands, a more powerful dynamo was fitted together with a new voltage regulator.

PRESS REACTIONS TO THE MK II

Relatively few road tests of the Mk II 3-litre Saloons were published in Britain, no doubt because editors and testers alike generally preferred to get their hands on an example of the more rakish Coupé. However, *The Motor* tested an overdrive model for its issue of 3 October 1962 (the car was 682 DNX, car number 770-00001a); *Country Life* of 1 November that year offered some road impressions of an unidentified overdrive car; and just over two years later, on 13 November 1964, *Autocar* published its test of an automatic model (registered BXC 629B).

The Motor test summarized the Mk II model eloquently:

> The image of the Rover as a comfortable means of transport for the elderly is no longer accurate. Nevertheless, with the high-quality execution of a design which lays stress on silence, luxury and roominess it consolidates its place as de luxe, executive-style transport.

The handling was much improved over earlier 3-litres, and although there was still some body roll, cornering was good for such a large and softly-sprung car. Not all the testers liked the loss of road feel which came with the power-assisted steering, and the report complained of long clutch travel and a ponderous gearchange, even though the remote lever was an improvement over the earlier type.

The car also displayed a number of faults which were not typical of the Mk II, having a spongy brake pedal and inconsistent servo operation; worse, the performance-test standing-starts caused first gear to 'fail', as the report delicately put it. In mitigation, however, it must be said that the car lent to *Autocar* was actually an Experimental Department test vehicle (which even had a development engine, number 3L/7B/54) and had probably already taken more than its fair share of punishment! It is likely that Rover was keen to get press coverage of the new Mk II before the Earls Court Show opened on 17 October that the press demonstrators had not yet been prepared, and that the only car available to lend to *Autocar* was 682 DNX.

John Eason Gibson, writing in *Country Life,* believed that the Mk II Saloon 'continues to cater for those who consider comfort, good engineering and good taste as the essential qualities in a car,' and that 'a wider public will be attracted by the higher performance now provided.' He went on to summarize the performance and handling improvements neatly:

> Broadly speaking, the improvements are such that the previous model's maximum speed – 98-100mph – can be regarded as the new car's cruising speed, and the previous model's optimum cornering speed in the hands of a ruthless driver can now be enjoyed by an average driver with even an apprehensive passenger on board.

Finally, the 3-litre made a first-rate long-distance car.

The longer the trip one does at one sitting the more impressed one is with the new 3-litre. Its qualities are such that, as well as finishing a long journey fresh and untired, one will often benefit from the relaxation possible and be sorry when one's destination is reached. There are few cars of which this can be said.

It was the quality of build and finish, together with the 3-litre's relaxed mile-eating ability, which most impressed *Autocar*'s testers in 1964. They wrote:

> Rover have achieved in a quantity-produced car a degree of finish, comfort and silence of which even the one-off craftsmen would have been proud, Quiet, smooth and comfortable, the 3-litre Rover with Borg-Warner automatic transmission as tested is one of the most relaxing cars in traffic that we have yet driven. On the open road it covers the miles in an equally relaxed, long-legged manner. It is beautifully finished, and it gives a lasting impression, however rough the road, of great strength and rigidity.

The mean maximum speed achieved on test was only 102mph, which did not compare favourably with the 108mph that *The Motor* had achieved two years earlier with an overdrive Mk II – but with near-silent cruising right up to 100mph and beyond, the car undeniably exhibited the blend of qualities expected of a luxury saloon in the mid-sixties.

RIVALS

Rover considered that the improved specification of the Mk II 3-litres justified higher showroom prices, and the 1963-season cars announced at the Earls Court Motor Show in October 1962 were rather more than 5 per cent dearer than the last of the Mk IA models. This moved them well out of range of the cheaper Humbers and Vanden Plas Princess which had been snapping at the 3-litre's heels during the 1962 season, and left them firmly in Jaguar territory. Jaguar had meanwhile hit back with the new Daimler 2½-litre V8, based on the compact Jaguar saloons and priced just below the 3-litres; but the Daimler was a much less spacious car than the Rover despite its good refinement.

The Jaguar counterattack was even stronger for 1964, because the October 1963 Earls Court Show saw the introduction of the new S-type models. The less powerful 3.4-litre was priced to compete directly with the Rover, although the more powerful 3.8-litre Jaguar was rather more expensive. Both had less interior room but a lot more performance than the Mk II 3-litre, but it was quite clear from their specification that they were aimed at the luxury saloon market which the Rover still dominated, rather than at the more overtly sporting customers who formed Jaguar's traditional market.

Humber and BMC could not match this twin onslaught. Both companies were struggling to sell their admittedly well-specified cars at prices considerably cheaper than the Rovers and Jaguars, and both appear to have concluded that the only way to gather more sales was to improve specifications and price their cars above the 3-litres and the S-types. As a result the October 1964 Motor Show saw the deletion of the Vanden Plas Princess 3-litre and its replacement by the very much more expensive Princess 4-litre R with its Rolls-Royce engine, while Humber added a new Imperial model derived from its Super Snipe and selling at a premium over the Rover. The position remained the same at the October 1965 Show, when the Mk II 3-litres made their final appearance alongside the new Mk III models.

PRODUCTION CHANGES AND THE SUFFIX 'b' AND SUFFIX 'c' CARS

The three-and-a-half seasons of Mk II 3-litre production saw a number of minor specification changes, and these are detailed in the table. The more major changes were of course accompanied by changes in the type suffix letter, from the 'a' of the first cars, through 'b' and ending with 'c'.

The introduction of the 'b' suffix in approximately December 1962 resulted from a change to the steering box of 3-litres without power-assisted steering. The original 17.6:1 ratio box was changed for a new Burman F3 type with a higher 20.3:1 ratio, which made the steering lighter and more manageable at parking speeds. Damping on the older system had been by friction discs in the steering relay, but now the undamped relay used on the power-steered cars was fitted, together with a separate hydraulic damper. (*Automobile Engineer* magazine published an interesting technical article on the developments leading up to this steering change in its January 1963 issue.)

Various running changes affected the 'b' suffix cars subsequently, but the two most obvious ones were made some time around September 1963 for the 1964 model-year (the precise dates are not clear). The first of these was the replacement of the chromed fuel filler cap by a flush-fitting cover panel painted in the body colour and locked by a catch inside the boot. The second was the relocation of the windscreen-washer jets in the bonnet panel. The chromed air intake panel nevertheless retained the two ribs on which the earlier jets had been fitted, and would do so until the very end of P5 production more than ten years later.

The change to the 'c' suffix specification occurred in March 1964, and marked the introduction of a modified engine and some

Rover 3-litre Mk II Saloon and Coupé: production changes

Note: All dates given for changes where no chassis number is quoted must be treated as approximate. The dates are taken from issues of *Rover Service Newsletter*, which usually reported changes between one and four months after they had actually been made on the assembly lines. Dates given for changes where a chassis number is quoted are exact for home market models and reflect the date into Despatch (i.e. the date the car was transferred from the assembly lines to the Despatch Department) given in records held by the British Motor Industry Heritage Trust.

Date	Home, Manual	Home, Auto	Other	Remarks
September 1962	Mk II Saloon and Coupé models introduced.			
		775-00124a	776-00052a 778-00003a	Nylon ball joints replaced rods and clips on automatic gear selector; retro-fit possible.
November 1962				Avon Turbospeed Mk IV tyres now fitted as alternative to Dunlop Roadspeed RS5 type.
December 1962	'b'-suffix Saloons introduced; Coupés retained 'a'-suffix.			
				Lower ratio (type F3) manual steering box with hydraulic steering damper; PAS steering relay commonized with manual steering cars; retro-fit possible.
January 1963				Longer control tube for gear lever reverse stop, and slotted nut instead of hexagon nut fixing upper gear lever to lower.
February 1963	770-00782b	775-00949b	771-00083b 773-00107b 776-00307b 778-00115b	Body-colour filler cap introduced to replace chromed type.
	735-00046a	740-00049a	738-00013a 743-00006a	
March 1963				Higher up-change speeds for automatic transmission.
May 1963				Rubber sleeve fitted between gear lever and knob to prevent buzz, from gearbox no. 770-01150a.
June 1963				Oil recommendation for overdrive changed from 90 EP to SAE 20 to prevent sticking when cold.
				Serrated wheel studs introduced and front hub flange widened; retro-fit possible.
July 1963	770-01531b	775-01895b	771-00148b	High-compression dampers

735-00176a	740-00140a	773-00312b 776-00427b 778-00311b 736-00009a 738-00036a 741-00014a 743-00015a	introduced, with 25 per cent increase in blow-off pressures, to prevent rear suspension bottoming in a fully-laden car and to improve body control at high speeds; retro-fit possible.

1964 model-year

September 1963	775-02390b 740-00298a	776-00487b 778-00440b 741-00020a 743-00064a	Nylon ball joints added to control rod between gear selection lever and compensator on automatic models; retro-fit possible.
December 1963			Lodge HLN sparking plugs approved as alternatives to Champion N5 with all 8:1 and 8.75:1 compression engines.
			Dynamo ratio changed from 1.5:1 to 1.8:1, by means of new dynamo pulley on all cars, plus new jockey pulley on cars with manual steering; from engine numbers 770-02948, 771-00485 and 775-03953. (This change to prevent contact burning in regulator.)
			Brass rings deleted from rubber boots on brake callipers, and pistons modified to suit; retro-fit possible.
January 1964			Modified wheel nuts to prevent damage to wheel discs, from axles numbered 770-03665A and 775-04148A.
			Lightweight grille with assembled grille rib frames in place of individual ribs; retro-fit possible.
			Pillar pulls fitted as standard to Saloons.
			Plastic coat hooks replaced metal type on Saloons and Coupés.
February 1964			Modified fan pulley and crankshaft vibration damper assembly with softer damper flywheel for smoother running, for manual-transmission cars only, from engines numbered 770-02708a and 771-00398a.

Core holes deleted from oil pump cover casting to prevent fractures, from engines numbered 770-01608a, 771-00306a and 775-02327a; retro-fit possible.

Lockers deleted from flywheel fixing bolts and bolt tightening torque increased.

New clutch driven plate with softer springs to prevent snatch and vibration, from engines numbered 770-01352a and 771-00276a.

Rover 2000 type courtesy light switches fitted to door pillars (with drive screw retention).

New front door ventilator window assembly on Coupé models, with improved sealing between glass and rubber at rear edge, improved entry of frame into rubber, and modified pivot pin to prevent damage to locating bush teeth; retro-fit possible.

March 1964   'c'-suffix models introduced.

Larger diameter crankshaft main journals; oil drain added to rear of cylinder head; new flywheel housing, inlet manifold and accelerator controls; new oil level unit in sump; engine lifting brackets with round holes instead of slots.

Revised instrument panel; speedometer with 25 per cent larger numerals and Arabic instead of Roman numerals for gear change markings; yellow minimum oil level marking added to fuel gauge; combined oil level and petrol reserve switch.

Two-speed wiper motor with rheostatic control switch.

Pointer replaced ring on gear selector of automatic models, and painted groove added to indicator plate.

Improved cold start control switch.

New Lucas type 90 SA switches for headlamps, overdrive control (manual models) and intermediate speed hold (automatic models).

Larger fuse box, now containing all fuses.

New wiring harness to suit electrical modifications.

Heated rear window with larger heated area; retro-fit possible.

Anchor points for safety harness fitted as standard; minor trim modifications to suit.

Sealed ball joints in front suspension.

Simplified centre exhaust mounting and modified third mounting, to allow increased fore and aft movement.

Redesigned main and reserve petrol pipes for better sealing.

Aluminized steel rear exhaust silencer.

Radiator now mounted to body only instead of to body and sub-frame, to prevent stress.

Redesigned battery box with improved dust sealing.

Blanking plates added to rear overriders for certain export territories.

Improved outside door handles with third fixing below push-button; longer button travel on Coupé models.

Tyre pump deleted from tool kit.

March 1964

Stronger differential unit with new material specification, from axle suffix 'b'.

	Brass stud and nut on front wing moulding finisher, to prevent rusting; retro-fit possible.
April 1964	New overdrive solenoid with adjustable stop, from gearbox number 770-04139a.
	Strengthening plate added to automatic gear change support strap.
	Girling Mk IIA brake servo replaced Mk II (identifiable by smaller air filter with screw fixing).
	Nylon pipe fitted between carburettor and fuel filter bowl.
July 1964	Damping ring deleted from front brake discs and anti-squeal shims added to brake pads to compensate.
	Grey steering wheels and steering column shrouds no longer available.
1965 model-year	
September 1964	Duaflex scraper rings replaced Maxilite type, to improve oil consumption; retro-fit possible.
December 1964	Sachet of Clearalex screen wash additive included with literature packs on new cars.
	Spherical-shaped rubber bush for radius arm to prevent failure from cutting and chafing.
	Improved overdrive with larger accumulator and stronger springs, from gearbox number 770-05682a.
	Neoprene washer added to nylon ball joint at upper end of automatic gear selector control rod, to prevent rattles and exclude dirt; retro-fit possible on Mk II models.
April 1965	Brass joint washer added to differential pinion housing, to form an oil reservoir and increase lubrication of the bearing.

June 1965	Flat dipping rear view mirror made optional (this was already standard on cars for the North American Dollar Area).
1966 model-year	
November 1965	Fitted floor mat sets optionally available, in charcoal grey rubber or beige nylon fur.
December 1965	Final Mk II models built.

electrical revisions, along with sealed front suspension ball joints, exhaust changes and other minor modifications. A prototype 1964-season 3-litre Automatic built on the production lines around May 1963 was designated a 'c' suffix car, which suggests that the original plan was to have the introduction of the 'c' suffix cars coinciding with the start of the 1964 model-year; however, it was not to be.

The modified engine (which, rather confusingly, had a 'b' suffix) had crankshaft main journals of a larger diameter than earlier types, designed to overcome a torsional vibration of the shaft encountered with the Weslake head. The castings of both block and cylinder head were also extended to provide an oil drain from the cylinder head into the sump, and this and other minor differences made interchangeability of components problematical between these engines and earlier types.

The electrical revisions included a new fuse box with four fuses to cover all the major circuits, so that there was no longer any need for an in-line fuse holder to cover the overdrive circuit on manual-transmission cars. (Nevertheless, the optional radio always would have a separate in-line fuse.) Two-speed wipers were added, with a rheostatic control switch which took the place of the fuel

An anecdote from the Development Section

Rover policy was to test modifications thoroughly, but very occasionally something slipped through the net. Project Engineer Dan Clayton remembers that,

Early P5s had chrome-plated coat hangers over the rear doors, until someone pointed out that they were projections that could cause serious injury to a rear passenger in the event of a rollover. Styling came up with a neat answer in the form of an oversize rear collar-stud-shaped nylon moulding, which was promptly adopted and brought into production. It looked fine, but unfortunately proved to be worthless for its intended purpose, as the suspended coat or hanger rolled off the new hanger when the car went round a corner! I recall being rebuked for not checking its performance before releasing it for production, and (I think deservedly) felt I had been treated rather unfairly – who could ever imagine the need to test or develop a coat hanger, for goodness' sake?

91

Mk II 3-litre Saloon: commission numbers and production figures

1963 Mk II Saloon	770-00001 to -01682	Manual, Home Market	(1,682)
	771-00001 to -00173	Manual, Export RHD	(173)
	772-00001 to -00156	Manual, CKD RHD	(156)
	773-00001 to -00362	Manual, Export LHD	(362)
	775-00001 to -02095	Auto, Home Market	(2,095)
	776-00001 to -00042	Auto, Export RHD	(42)
	777-00001 to -00042	Auto, CKD RHD	(42)
	778-00001 to -00356	Auto, Export LHD	(356)
			Total: 4,908
1964 Mk II Saloon	770-01683 to -03238	Manual, Home Market	(1,556)
	771-00174 to -00347	Manual, Export RHD	(174)
	772-00157 to -00216	Manual, CKD RHD	(60)
	773-00363 to -00566	Manual, Export LHD	(204)
	775-02096 to -04293	Auto, Home Market	(2,198)
	776-00043 to -00779	Auto, Export RHD	(737)
	777-00043 to -00114	Auto, CKD RHD	(72)
	778-00357 to -00612	Auto, Export LHD	(256)
			Total: 5,257
1965 Mk II Saloon	770-03239 to -04711	Manual, Home Market	(1,473)
	771-00348 to -00527	Manual, Export RHD	(180)
	772-00217 to -00252	Manual, CKD RHD	(36)
	773-00567 to -00730	Manual, Export LHD	(164)
	775-04294 to -06802	Auto, Home Market	(2,509)
	776-00780 to -00815	Auto, Export RHD	(36)
	777-00115 to -00150	Auto, CKD RHD	(36)
	778-00613 to -00771	Auto, Export LHD	(159)
			Total: 4,593
1966 Mk II Saloon	770-04712 to -04857	Manual, Home Market	(146)
	771-00528 to -00554	Manual, Export RHD	(27)
	772-series (none)	Manual, CKD RHD	(–)
	773-00731 to -00751	Manual, Export LHD	(21)
	775-06803 to -07099	Auto, Home Market	(297)
	776-series (none)	Auto, Export RHD	(–)
	777-series (none)	Auto, CKD RHD	(–)
	778-00772 to -00799	Auto, Export LHD	(28)
			Total: 519

Total: 15,277

pump changeover switch on the instrument panel. In turn, the oil level indicator switch between the two main instrument dials was made to double as a reserve fuel pump switch, and the markings on the face of the instrument panel were changed to suit. The instrument panel itself also changed, and the 'c' suffix Mk II had a panel moulded from matt black plastic, with a black leathercloth facing in its recessed centre section. The speedometer, too, was changed, and on the 'c' suffix cars it had larger markings and Arabic numerals for the maximum gearchange speeds. The fuel gauge on the combined-instruments dial, meanwhile, now had a yellow marking for the oil-level indicator.

No major changes were made for the last full season of Mk II 3-litre production, beginning in September 1964, although power-assisted steering was made standard. Rover Company publicity claimed that more than a quarter of 3-litre Saloons were being ordered with power-assisted steering by mid-1964, and of course all Coupé models had it as standard. So this change must have streamlined production, as well as enhancing the standard specification of the car.

THE MK II IN THE USA

There was a feeling at Rover that the new P6 2000 Saloon, scheduled for introduction in autumn 1963, stood a much better chance of attracting sales in the USA than any previous model from Solihull. It was an altogether more modern and less staid car than existing Rover products, and the company wanted to make sure the 2000 was given the best support possible in the USA. So, early in 1963, the Rover Motor Company of North America (RMCNA) was revitalized with new headquarters in New York at 405 Lexington Avenue, and a new management team was appointed. At its head was Bruce McWilliams, who had formerly run the Mercedes-Benz import operation under Studebaker.

The 3-litre did not sell well in the USA, but that did not prevent Rover from having a special US specification for the Mk II cars. The whitewall tyres were only to be expected, but that front wing badging was unique to US market cars.

McWilliams was an energetic and enthusiastic individual, and he set about improving Rover's image – and, more importantly, its sales – in the USA and Canada. Some of the Rover publicity of the time, for which McWilliams' English-born wife was responsible, was quite unforgettable. One well-known advertisement for the Land Rover made capital out of its use as a getaway vehicle in the Great Train Robbery of August 1963, while sales brochures in the shape of Easter eggs were used to promote the 3-litre Saloon and Coupé at Easter that year.

It would probably be fair to say that the focus of RMCNA's promotional activity switched to the new 2000 when it became available in the USA during 1964, and that what little interest the company had aroused in the 3-litres waned after that. Nevertheless, both Saloon and Coupé models remained available until the end of the Mk II production run, characterized by whitewall tyres as standard and by a chromed 'Rover' badge ahead of the trim strip on each front wing. That badge echoed the style originally proposed for the first 3-litres in 1958 but actually used only on the works rally team cars in 1963–1964.

One problem which McWilliams inherited was a stock of unsold cars, with the result that the February 1963 road test of a 3-litre by *Car and Driver* magazine was actually of an old-stock Mk IA Saloon. The magazine reported fairly favourably on the car, but of course the Mk II would have created a more favourable impression. When the Mk IIs did become available to RMCNA, the company lost no time in trying to persuade the leading magazines of the day to borrow examples for road test. However, in a country where major annual facelifts were the norm in the automotive industry, the public wanted to hear about revolutionary changes, not evolutionary ones. So the Mk II Saloons were ignored by the press, and

> **Mk II models for the Royal Household**
>
> The Royal Family took delivery of at least two Mk II 3-litre Saloons. The first one, delivered in or around March 1963 for the personal use of Her Majesty the Queen, was registered as JGY 280 – the number of the original Royal Mk I. The second car was built rather later and was finished in what appears to be Pine Green. This later car is now on show at the Royal Mews in Sandringham.

probably the only reason why *Road and Track* agreed to try out a Coupé (*see* Chapter 5) was that it did at least look different from what had gone before.

THE CKD MK II 3-LITRES

The South African CKD operation, which had started in 1961 with manual-transmission

More overseas oddities: these seats are in a South African-assembled Mk II Saloon. Lyback reclining seats were never available in that market, and even the standard individual front seats lacked the armrests found on Solihull-built cars.

Mk IA models, continued during the period of Mk II production. Both manual and automatic Mk II Saloons were shipped to South Africa in CKD form between 1962 and 1965, for local assembly. These had a few differences from their UK-assembled equivalents, no doubt because some components were manufactured locally in South Africa. There were minor pedal differences, possibly a few switchgear differences, and the front seats were invariably fixed-backrest individual types without the pull-down armrests found on UK-built cars.

There were no CKD Coupés, and the South African operation ceased before the Mk III models came on-stream in the middle of 1965. A total of 402 cars (252 with manual transmission and 150 automatics) were involved.

THE MK II 2.4-LITRE CARS

There is no hard evidence that Rover ever built any Mk II 2.4-litre cars, and neither the factory's despatch records nor any actual cars have yet come to light. Nevertheless, it seems beyond doubt that such a model was at least planned, and references in Rover service documents strongly suggest that examples may have been built in quantity.

The second edition of Rover's parts list for the 2.4-litre and 2.6-litre cars, dated February 1965, refers to Mk II 2.4-litres and actually lists a Mk II 2.4-litre engine and flywheel assembly with a separate part number from the Mk IA type. There is also a list of Rover and Land Rover commission numbers dated December 1965 which shows that the following commencing numbers had been allocated to the cars:

790-00001a	RHD manual, home market
791-00001a	RHD manual, export
792-00001a	RHD manual, CKD
793-00001a	LHD manual, export
794-00001a	LHD manual, CKD

Commission number plates had almost certainly been made as well, because one was illustrated by mistake in the January 1963 *Workshop Manual* which Rover produced for the P4 models: the illustrator must have been given the wrong plate to copy for his picture of a 110 chassis plate, and his picture clearly shows a plate reading 'Rover 2.4-litre Mk II'!

Rover 2.6-litre Saloon (1962–1965)

Specification as for contemporary 3-litre with manual transmission, except:

Engine

Type	Short-stroke 3L7 with Weslake head
Block material	Cast iron
Head material	Aluminium alloy
Cylinders	Six, in line
Cooling	Water
Bore and stroke	77.8mm × 92.075mm
Capacity	2,625cc
Main bearings	Seven
Valves	Inlet valves in cylinder head and exhaust valves in cylinder block
Compression ratio	8.8:1
Carburettor	Single SU, type not known
Max. power	123bhp at 5,000rpm
Max. torque	142lb.ft at 3,000rpm

Transmission
All models had manual transmission, but it is not clear whether overdrive was standard or not.

Steering
It is not clear whether power-assisted steering was ever fitted to these cars.

Mk II 2.6-litre Saloon: commission numbers and production figures			
1963 Mk II	781-00001 to -00006	Manual, export RHD	(6)
	783-00001 to -00043	Manual, export LHD	(43)
			Total: 49
1964 Mk II	781-series (none)	Manual, export RHD	(–)
	783-00044 to -00070	Manual, export LHD	(27)
			Total: 27
1965 Mk II	781-00007 to -00014	Manual, export RHD	(8)
	783-00071 to -00092	Manual, export LHD	(22)
			Total: 30
			Total: 106

THE MK II 2.6-LITRE CARS

There is no question mark over the existence of Mk II 2.6-litre models, however. These cars were identical in most respects to their 3-litre contemporaries, but they had 2.6-litre engines and special badging. Like the Mk IA versions, they carried a 'Rover 2.6-litre' grille badge, while the boot lid carried a chrome script badge like that on the 3-litre which read '2.6 litre'. All the 2.6-litre cars had manual transmissions, but it is not clear whether all of them were fitted with overdrive; nor is it clear whether power-assisted steering was ever fitted.

These 2.6-litre cars were built in both left-hand and right-hand drive forms. All ninety-two left-hand drive cars, plus one right-hand drive example, went to Franco-Britannic Autos in Paris. The remaining thirteen right-hand drive cars were shipped to the British West Africa Corporation in Lagos, Nigeria. All the 2.6-litre cars were Saloons, even though the Rover Despatch Department recorded them as Coupés.

The P5 Project team

The engineer responsible for co-ordinating the P5 Project before 1964 was Dan Clayton, but in that year he was moved to a new job. Rover advertised the vacancy, and attracted a number of applicants. The man who succeeded Clayton was Ken Stansbury, who had been Chief Development Engineer at Armstrong-Siddeley before moving to a job at Dunlop, where he worked on brakes and suspensions among other things. Stansbury remained P5 Project Engineer from 1964 until the last P5 was built in 1973, and for most of that period was supported by Alan Tudor as Assistant Project Engineer. At any one time, there would have been one or more Technical Assistants also working on the team.

Stansbury remembers that he was short-listed for the Rover job with Geoff Hills, who had worked on the Humber Super Snipe. Hills also took a job at Rover – as Project Engineer in charge of the P6B. However, it is clear from the backgrounds of the two men – in Armstrong-Siddeley and Humber respectively – that Rover particularly wanted someone who had experience of big luxury cars.

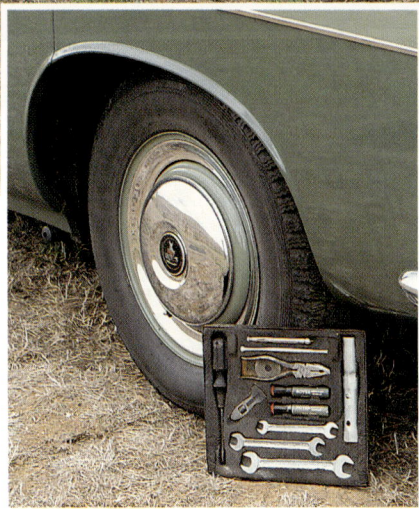

(Above) *The classic lines which would serve the big Rover so well for fifteen years are clear in this picture of Peter Harrison's 1960 Mk I 3-litre saloon. The car has lost its original 'tripod' headlamps and is running on radial tyres, but looks otherwise as it would have done when new.*

(Inset) *Mk I cars had chromed steel hubcaps with a separate outer trim ring. Neither resisted corrosion very well. The tool tray contains touch-up pencils for the car's two-tone paintwork.*

The beautifully presented engine bay of Trevor Hallett's 1961 Mk I clearly shows the original oil-bath air filter.

(Above) *The Coupé carried additional instruments in pods slung under the main binnacle. The facia of this overdrive model is otherwise identical to that of the Mk II Saloon, with black steering wheel and column shroud, and much less chrome on the instrument binnacle than was found on earlier cars.*

The front seats in the 3½-litre Coupé were as comfortable as they look; these are in Saddle Tan. The tool tray pulled out from the centre of the dash – but the transmission selector had to be pulled back out of the way first. The tray had a laminated plastic cover in imitation wood veneer, which has been removed for this photograph.

(Left) Built-in foglamps, rubber-faced bumper overriders and a gilded grille badge characterized the V8-engined cars. Note that these cars also had no provision for a starting-handle.

Patrick Swan's 1965 3-litre Coupé wears the optional fog and spot lamps on its front bumper. The car is finished in Steel Blue with a Charcoal roof. The starting-handle guide is visible at the base of the grille.

Chromed Rostyle steel wheels were fitted to the V8-engined cars. All had centre caps with a gold Viking ship emblem, which quickly turned white when exposed to sunlight.

(Below) The lowered roofline and large rear window of the Coupé are evident in this view. The parcels shelf of Coupés was trimmed in leathercloth to match the seats; Mk II Saloons still had a self-coloured black shelf with an indentation.

The coachline and the circular chromed filler cap are clear in this picture of a 3½-litre Saloon's rear end.

Coupés always had individual rear seats; these are in a 3½-litre model. The car is fitted with a single front headrest on the passenger's side, and the rear seat passenger's reading lamp can be seen underneath it.

Plate badges were fitted to the V8-engined models, and there were side repeater flashers. The cars also had a painted coachline below the bodyside trim strip.

(Left) These two 3½-litre cars were lined up on Westminster Bridge early one morning in 1988 for Classic and Sportscar *magazine, with whose permission this picture is used. The Coupé in the foreground has a Silver Birch roof over a Zircon Blue lower body. The Saloon behind is finished in Bordeaux Red, and carries the optional badge bar.*

The door trims of this Coupé show the style adopted for Mk II models. Also visible are the stainless steel door tops and the wooden door cappings which were both unique to the Coupé.

This is the broad bench rear seat of Peter Harrison's Mk I. The rear parcels shelf has a practical indentation to prevent objects sliding off it, and the louvre over the drop-glass is clearly visible.

(Above) *This is the Mk II engine bay, actually in a Coupé as the twin inspection lamps under the bonnet make clear. The twin-trumpet air cleaner has lost some of its paint, probably through the removal of the stickers applied during routine maintenance by garages. Alongside the washer bottle on the far side of the engine bay is the fluid reservoir for the power steering which was standard on Coupés.*

The double-barrelled exhaust fitted to the V8-engined models peeps out from beneath the bumper of this Coupé.

5 The Coupé – Origins and Mk II Models

The first Coupé prototype had frameless door glass, as did the Rover 2000 prototypes built around the same time. Problems with the frameless windows led to a redesign, in which both cars were given thin stainless-steel window frames. This is the frameless-window Sportsman prototype pictured in July 1959. The car was clearly new and unregistered when this picture was taken, and the date suggests that this was actually the second prototype.

Maurice Wilks' idea for a second variant of the P5 range began to take shape in the first half of 1957. Among the first things to happen was that David Bache was asked to sketch up some proposals for this new model, which was at that stage always referred to as the Hard Top in imitation of American practice. The idea was to modify the car only from the waist up – much as the earlier Rover Sports Saloons had used a four-window passenger cabin instead of the six-window type on standard Saloons.

Bache delegated the job to Tony Poole, asking him to put forward and illustrate about half a dozen proposals. As Poole remembers it, 'Bache said to me, 'Now do two at the bottom that are really way out.' He says, 'Lots of curved glass.' So I did these two. And we knew, you know, 'No way!'" Much to the surprise of the whole Styling Department, it was

one of the two way-out sketches which the Wilks brothers chose when they came down to the Styling Department to review progress. 'We could have fallen through the floor!' remembers Poole, adding that the favoured proposal was quickly nicknamed the Dan Dare roof in honour of the hero of a science-fiction strip-cartoon in the contemporary *Eagle* comic.

As Maurice Wilks wanted the Hard Top to enter production alongside the standard Saloon, he lost no time in forging ahead. By the beginning of May 1957, he was planning to ask Pressed Steel to build some prototype bodies, the first to be delivered to the Engineering Department by February 1958 and the next to be available in time for a complete car to be displayed on the Rover stand alongside the Saloon at the 1958 Motor Show. Then, on 3 June he got the endorsement of his colleagues on the Rover Board to spend £150,000 on body dies for the new model. Confusingly, the minutes of that Board meeting refer to the car as a 'P5 'Hard Top Sports Saloon"; the car would be variously referred to as the 'Hard Top', 'Sports Saloon', 'Sportsman' and even the '3-litre S' before the definitive name of 'Coupé' was chosen some five years later!

Meanwhile, the new roof was built up as a full-size wooden mock-up in the Styling Department at Solihull, and by midsummer 1957, there was a mock-up at Pressed Steel. Jim Dodsworth was appointed as Project Engineer to co-ordinate the work of all those who had input to the new model's design and development, and he remembers that one of the early problems was with the instruments reflecting in the large curved windscreen. To cure this problem, a lip was added to the top edge of the instrument binnacle, and the lip was standardized for Saloon production as well.

Things were well on course by January 1958, when a pair of wooden quarter-scale models of the two projected P5 variants were taken out to Spencer Wilks' house at Monks Kirby to be photographed. The two models

---

The Coupé prototypes

Full details of the prototype Coupés have not survived, but existing Rover records held by the British Motor Industry Heritage Trust (BMIHT) and the memories of engineers who knew the cars have allowed the jigsaw to be partially reconstructed.

Pillarless prototypes

There were probably two of these, built in December 1958 and July 1959 respectively. One car was photographed extensively in July 1959, and it would be logical to assume that this was the second prototype.

Test engineer Brian Terry remembers one of these cars as being painted grey and white. It started life with automatic transmission but was later fitted with an overdrive manual gearbox. He believes its registration number contained the number 3990. BMIHT records show the best candidate for this car as P5/16 (the sixteenth P5 prototype) which was registered as 3990 NX on 21 January 1960. That date is rather later than the build dates of the two known cars, but Rover prototypes often did run on trade plates for several months before they were road-registered. 3990 NX had an early prototype or development engine, number 3L/7B/4 ('3-litre, seven-bearing, 'B' or Weslake head specification, number 4'). It was scrapped by Rover on 26 April 1962 – shortly after the first off-tools prototype had been completed.

Brian Terry also remembers that one of the early prototypes had a manual gearbox without overdrive and a 3.9:1 axle.

Off-tools prototypes

Although three cars were planned, details of only two are known. These are P5/48/CP1, registered as 464 CNX on 1 April 1962, and P5/50/CP2, registered as 693 DNX on 1 August 1962. 464 CNX had engine number 3L/7B/23, and 693 DNX had engine number 3L/7B/49A. It is curious that colour photographs of 464 DNX were airbrushed and used on the cover of a Mk III sales catalogue some three years later; surely Rover could have provided their printers with some more up-to-date pictures?

Brian Terry remembers that the maximum-speed testing on the Coupé was done with an overdrive car. It is likely to have been one of these two.

were placed on a board and provided with a realistic scale wall as background, and the real trees in the grounds at Wilks' house helped to produce some very realistic-looking pictures. The Hard Top model shows very clearly how the car was then conceived. It was to have frameless door windows in the American hardtop tradition, a reverse curve at the lower trailing edge of the rear door window, and an indicator repeater lamp (or possibly a parking lamp) on the textured steel plate behind that window. Two-tone paint was also obviously considered essential, as the model had a dark roof and a lighter coloured body. Lastly, the wheel trims (like those on the quarter-scale Saloon) had a painted trim ring which matched the upper body colour. This, according to Tony Poole, was a deliberate echo of Mercedes-Benz practice.

Pressed Steel were expected to deliver three prototype Hard Top bodies to Solihull by March 1958 so that the first engineering

prototype would be ready in July, but these were no doubt held up by the same inter-union dispute which affected delivery of the first off-tools Saloon bodies. As the whole P5 Project was now seriously behind schedule, the engineering policy meeting on 1st May decided that the Hard Top would now definitely not go to the Earls Court Show in September. Work on the project was therefore relegated to a lower priority while Solihull tried frantically to get the basic Saloon model ready in time for that show.

The urgency had gone out of the Hard Top project, but it was not long before work resumed on it. By mid-May, there was talk of the first engineering prototype being on the road during August, but further delays meant that the first prototype was actually not completed until just before Christmas 1958. By then, the plan was that the three engineering prototypes should be followed by three off-tools prototypes before production began.

Spot the deliberate mistakes! This press picture of the 3-litre Coupé was issued in the autumn of 1962, but it actually shows a well-used prototype with a number of differences from the production cars. 464 CNX was the first 'production-standard' car, had number P5/481/CP1 on the engineering fleet, and was registered on 1 April 1962. Note the Lucas 'tripod' headlamps (associated with Mk I and Mk IA models), the position of the catch on the vent window (it was at the bottom on production cars), and the absence of the Viking ship logo on the rear pillar's decorative plate. There are other tell-tales, too: Rover would never have let a production car out with trim strips fitting as badly as these (look at the poor match between back door and wing), and the rubber bung is missing from the front jacking point.

This slightly later press picture shows a real production car, with all the correct details. Two-toning in which the darker colour was used on the roof tended to emphasize the shallow windows and make the car look fat. The colours on this one are probably Steel Blue (really a light grey) and White.

The second and third Hard Top prototypes were expected to be ready in July 1959, and Solihull was hoping that the car would now be ready in time for the start of the 1961 season in September 1960. The second car was completed as planned in July, but the third car may never have been built. Tests on the first prototype had already revealed a number of problems and the project was put on hold early in April while some redesign work was done.

THE PRODUCTION DESIGN, 1960

This proved to be the real watershed in the development of the model. One of the major problems was that the original pillarless design could not be made to work in practice. Project Engineer Jim Dodsworth remembers that it proved impossible to come up with a window winding mechanism which would guarantee the door windows always rose to the same height and at the same angle, and test engineer Brian Terry recalls that, even when raised, the windows used to flap at speed and would not seal properly. Pillarless construction and frameless door windows had also been features of the first prototypes of the P6 saloon, and Rover ran into exactly the same problems –

compounded by body weaknesses – with those cars during 1959.

So both the P6 and the Hard Top were redesigned with full B/C pillars and with slim bright metal window frames. The redesign took a long time: the first P6 prototypes with framed doors did not appear until July 1960 and the Hard Top project lagged behind them. It was not until 2 June 1960 that the Rover Board even approved a further £50,000 for what it now called the P5 Sports Saloon project – money which can only have been needed in connection with the redesign.

Initial optimism that the Sports Saloon could have been put into production towards the end of 1961 or early in 1962 soon proved ill-founded. In fact, a more realistic schedule was not in place until August 1961, when Pressed Steel were expected to deliver three prototype bodies to the new design during October and November. The first of these would be assembled into a running prototype by the end of December; the second would take to the road in mid-February 1962 and the third would be ready by the end of March. A pre-production batch of off-tools cars would be built in June and July, and cars would be delivered to the Sales Department in October 1962. Meanwhile, the winter of 1960–1961 saw work going ahead on the Weslake-head engine. The Rover Board

Rover 3-litre Coupé (1962–1966)

Layout

Monocoque bodyshell with front subframe bolted in place. Four-seater saloon, with front engine and rear wheel drive.

Engine

Type	3L7 with Weslake head
Block material	Cast iron
Head material	Aluminium alloy
Cylinders	Six, in line
Cooling	Water
Bore and stroke	77.8mm × 105mm
Capacity	2,995cc
Main bearings	Seven
Valves	Inlet valves in cylinder head and exhaust valves in cylinder block
Compression ratio	8.75:1 with manual transmission 8:1 with automatic transmission
Carburettor	Single SU type HD6 (2in)
Max. power	134bhp gross (121bhp net) at 5,000rpm with manual transmission 129bhp gross (116bhp net) at 4,750rpm with automatic transmission
Max. torque	169lb.ft at 1,750rpm with manual transmission 161lb.ft at 3,000rpm with automatic transmission

Transmission

Manual models	Hydraulically operated single dry plate clutch, 10in diameter
Automatic models	Torque converter

Internal gearbox ratios

Option 1 Four-speed manual with overdrive

	(Final drive 4.3:1)
Overdrive	0.77:1
Top	1.00:1
Third	1.27:1
Second	1.88:1
First	3.37:1
Reverse	2.96:1

Option 2 Three-speed automatic (Borg Warner type DG)

	(Final drive 3.9:1)
Top	1.00:1
Intermediate	1.43:1
First	2.30:1
Reverse	2.00:1

Suspension and steering	
Front	Independent, with wishbones, laminated torsion bar springs and anti-roll bar
Rear	Semi-floating axle with progressive-rate semi-elliptic leaf springs
Steering	Power-assisted Burman recirculating ball type with variable ratio and 17.6:1 ratio at straight ahead position
Tyres	6.70 × 15 crossply
Wheels	Five-stud disc type
Rim width	5in

Brakes	
Type	Discs at the front and drums at the rear, with servo assistance
Size	Disc diameter 10.75in Drum diameter 1in

Dimensions (in/mm)	
Track, front	55 (1,397)
Track, rear	56 (1,422)
Wheelbase	110.5 (2,807)
Overall length	186.5 (4,737)
Overall width	70 (1,778)
Overall height	56.75 (1,441)
Unladen weight	3,727lb (1,690kg) (overdrive) 3,741lb (1,697kg) (automatic)

had allocated £70,800 to this project on 18 November 1960, when the engine had been intended purely for the Sports Saloon.

Pressed Steel were late in delivering the prototype bodies, and the schedule slipped again. Fortunately, however, the slippage was not disastrous. The first prototype – originally planned for the end of December 1961 – had been built up by 1 April 1962. This car had chassis number P5/48/CP1, and the inclusion of those letters 'CP' suggests that the definitive production name of Coupé had been chosen by that date (the other numbers indicate that the car was the 48th P5 on the Engineering Department fleet). The second prototype existed by 1 August 1962; whether the third prototype was ever built is not clear. The redesigned cars proved

satisfactory, and Rover decided to announce the new Coupé model in September.

COUPÉ CHARACTERISTICS

From the outside, only the different passenger cabin and a discreet 'Coupé' script badge on the bootlid distinguished the 3-litre Coupé from its Saloon equivalent, but sharp eyes would have spotted the body-colour fuel filler cover as well. Like the Saloon, the Coupé was made available with two-tone colour schemes to suit the taste of the times, although it must be said that not all of these were very flattering. The colour split was different from that on the Saloons, and the whole of the Coupé's lower body was finished

The 3-litre Coupé looked particularly attractive from behind. This well-preserved example pictured in 1995 shows off the 'Coupé' script badge which replaced the '3 litre' badge on these models.

side windows. This stylized badge had originally been drawn up in the Styling Department for use on the Rover apprentices' headed notepaper, and was only adapted for the car when something was needed to replace the small lamps originally intended to go on the textured plate. The adaptation was done by a left-handed draughtsman in the Styling Department, and no-one noticed until it was too late that he had turned the design around. As a result, the stylized Viking ship badge has its flag pointing to the left, while the flags on all other Rover badges point to the right!

Mechanically, the car was exactly the same as the Mk II 3-litre Saloons which were announced at the same time, with their slightly lowered suspension, Weslake-head engine, and the option of overdrive or three-speed automatic transmission. However, Coupés came with the power-assisted steering as standard, which was an extra-cost option on Saloons. This was one reason why Coupés were actually heavier than their Saloon equivalents by some 90lb (41kg), a difference which more than offset the minor aerodynamic advantage of their lower profile. In fact, an automatic 3-litre Coupé was almost embarrassingly slower than an overdrive Mk II Saloon – a fact which might not have helped sales of the more expensive Coupé if it had been widely publicized at the time!

in one colour while only the roof was painted in the second colour. The car looked at its best with a dark lower body and light-coloured roof; reversing dark and light tones tended to emphasize the shallowness of the side windows and to make the car look rather fat. As Jim Dodsworth remembers, 'We were all very aware that those small windows made the Coupé look even more like a tank than the Saloon did!'

Tony Poole tells an amusing story about the small stainless-steel Viking ship badge affixed to the textured plate behind the rear

This is the body-coloured fuel filler cap introduced during 1963 on Coupés and to replace the original chromed type on Saloons. It was locked by a catch accessible only from inside the boot.

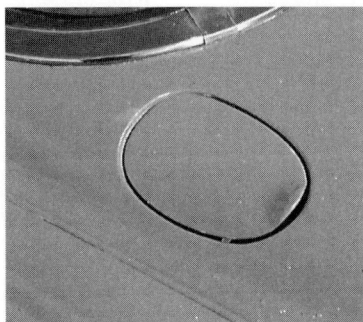

The instrument panel of the Coupé featured four minor gauges in pods underneath the main binnacle, while the large dial to the right of the speedometer was a rev counter, red-lined between 5,250 and 5,500rpm. This is a suffix 'c' car with manual transmission and overdrive.

(Left) The interior presented four individual seats, the front pair being the fully-reclining type. Rear legroom and headroom were slightly more restricted than in the Saloon, but the car was still a spacious and comfortable long-distance tourer.

Many of the Coupé's special features were of course inside the car. The front seats were the optional fully-adjustable 'Lyback' Saloon items, and in place of the rear bench came two individual seats, with a neat 'smoker's companion' of ashtray and cigarette lighter between them. Compensation for the 2½in of headroom lost to the lower roofline came in the shape of seat cushions mounted lower than the bench in the Saloons, and in more steeply raked seat backs. The net result was a slight loss of legroom in the rear.

The bottom of the B/C post on Coupés was shaped differently from its Saloon equivalent, mainly to give sufficient structural rigidity to the body. As a result, the inner door panels were also slightly different. They in turn carried unique door trims, similar in general design to the Saloon type but with window winders slightly relocated and a unique capping of African cherry wood. This was much deeper than the Saloon type to compensate for the lack of wood around the window frames, but was neatly mated with the existing dashboard furniture by a bright metal strip. Black plastic stalks emerging from these wooden cappings allowed the doors to be locked quickly and easily from the inside.

Also unique to the Coupés was the rear parcels shelf, trimmed in leathercloth to

---

Paint and trim colours, Mk II Coupé

The 3-litre Coupé was available in ten single-tone colours and in ten two-tone combinations. There were six different upholstery colours, in each case with carpets to match. The interior trim options with single-tone cars were the same as those for single-tone Saloons (*see* Chapter 5). The standard combinations for two-tone Coupés were:

Roof	Lower body	Upholstery
Black	Charcoal	Blue, Green, Grey, Red, Stone or Tan
Charcoal	Steel Blue	Blue, Grey or Red
Juniper Green	Pine Green	Green, Grey, Stone or Tan
Light Navy	Marine Grey	Blue, Grey, Stone or Tan
Marine Grey	Black	Blue, Green, Grey, Red, Stone or Tan
Pine Green	Stone Grey	Green, Stone or Tan
Steel Blue	Light Navy	Blue, Grey or Stone
Steel Blue	White	Blue, Grey, Red, Stone or Tan
Stone Grey	Burgundy	Grey, Red or Stone
Stone Grey	Juniper Green	Green, Grey, Stone or Tan

---

match the seats and flat rather than dished as in the Saloons. The main point of this redesign was to allow a second radio speaker to be mounted beneath the trim for the benefit of rear seat passengers. Up front, meanwhile, the instrument binnacle now contained a rev counter alongside the speedometer. The fuel gauge, ammeter and

From October 1963, the 2000 joined the Rover range alongside the 3-litres and the run-out P4 models. This picture was taken at that year's Frankfurt Motor Show, and shows a left-hand drive 2000 and the 2000 cutaway alongside a 3-litre Coupé equipped with the 'racing' mirrors favoured by the Germans. The narrow-band whitewalls set off the car's dark paint very well.

water temperature gauge displaced by this addition were relocated in small pods below the main binnacle, where they were joined by a matching oil-pressure gauge. The overall result of this redesign was to strengthen the Coupé's more sporty image – even though the reality of the car's performance was rather different.

Last but not least, the Coupé was the first Rover to have zero-torque door locks, which would also be introduced on the 2000 saloon

a year later in October 1963. The theory behind these was that the lock barrels could only be turned one way – the correct way – so that locking and unlocking were simplified. As the lock barrels could not be incorporated into the door handles in the way they had been on Saloons, Coupés always had plain press-buttons in their front door handles and the locks located just beneath the handles.

THE 3-LITRE COUPÉ ON SALE

Sales of the 3-litre Coupé were slow to get under way. Although the new model was announced in September 1962, only a tiny handful of cars actually existed by then. The first two production cars were both automatic models built at the beginning of September: 740-00002A was a Burgundy car finished on 6 September, a day earlier than 740-00001A which was two-toned in Pine and Juniper Green. Both became press vehicles, registered as 837 DWD and 838 DWD, respectively. The next car to be built was a White left-hand drive overdrive model for the Paris Show (738-00001A) which was completed on 27 September. Three more cars followed on 10 October, 1962, and were a Steel Blue and White automatic for the Earls Court Show (740-00003A), a Steel Blue and Light Navy overdrive car for the same show (735-00001A) and a further overdrive model (735-00002A). A third overdrive car for the home market (735-00003A) was built in November, but full production did not begin until January 1963.

Sales catalogues nevertheless had to be ready by the time of the Earls Court Show in October 1963, and the artist responsible for the pictures had to know what a 3-litre Coupé looked like. A careful look at the original sales catalogue for the Mk II models makes clear that the car he saw differed from production standard in some respects: his illustrations

---

Optional extras for the 3-litre Coupé

The 3-litre Coupé was a very well-equipped car for its day, but customers could order the following at extra cost:
Badge bar
Dipping interior mirror (from October 1965)
Electric immersion heater for cylinder block
Exhaust tail pipe finisher
Extension speaker for radio (fitted under rear parcels shelf)
Floor mats, beige nylon fur (from October 1965)
Floor mats, Charcoal Grey rubber (from October 1965)
Fog lamp
Foot pump (standard equipment before October 1964)
Heated rear window
Laminated windscreen
Mud flaps, front and rear (from October 1964)
Radio (Pye or Radiomobile)
Seat belts (Irvin) for front and rear
Special paint finish or interior trim to customer's requirements
Spot light
Towbar
Two-tone paint
Whitewall tyres
Wing mirrors

show a Coupé with quarter-light catches positioned half-way up the front window pillar and the Lucas tripod-type headlamps associated with the Mk IA and earlier cars. These features were also found on the first off-tools prototype, 464 DNX, which may well be the car he was shown.

That delayed start to sales ensured that relatively few Coupés were delivered before the end of the 1963 season. Only 417 found customers at home, while exports were – as expected – trifling. Overdrive cars proved more popular than automatics, which suggested that some customers at least believed Rover's sales pitch that the Coupé was a more sporting model than the Saloon. It became clear pretty soon that it had no more performance, however, and during 1964 the preference switched decisively to automatics. That year proved the best, with total sales reaching 2,503 cars. For 1965, sales dropped by around 16 per cent, and of course only a small number of Mk II Coupés left the lines during the 1966 season because the new Mk III models had become available.

The first proper road test of a 3-litre Coupé was published by *Autocar* in its 5 July 1963 issue, by which time the car had been available through the showrooms for some six months. The car in question was an overdrive model, 735-00019A, registered 586 EWD on 1 March 1963. It underperformed at first, and was returned to Rover for some attention before recording a 109mph top speed. For the *Autocar* staff, who had not tested a 3-litre of any kind since August 1959, the interest was as much in the changes made to the P5 over the previous four years as in the Coupé body configuration. However, they did comment that although the lowered roofline was matched by a lowered rear seat cushion so that headroom was not affected as compared to the Saloon, 'passengers now sit with their knees up and cannot see out quite as well.' They

also noted, with the restraint characteristic of the period, that this was no sports saloon:

Some cars, of which Rover is one, seem to emit an aura that manages to instil itself into the driver and passengers until they take on a measure of this character and reflect it in their driving and treatment of the car. Even the exuberance of youth would be quickly tempered after a brief spell at the wheel. Yet when a fast mood takes one, the car responds eagerly, and covers distances rapidly and with inconspicuous efficiency.

What, then was the appeal of the 3-litre Coupé? Rover had always planned it as a high-performance variant of the P5 range, and only the need to improve the Saloon's performance had caused both models to share the Weslake-head engine which had originally been planned for the Coupé alone. That vision of high performance remained in Rover advertising, and as late as December 1964 one magazine advertisement described the Coupé as 'the fastest car in the Rover range.' Perhaps it was, in terms of top speed, but the new P6 2000 saloon ran it a very close second, and the 2000's superior handling meant that a well-driven example would be able to cover the ground much more quickly.

A couple of other epithets from that 1964 advertisement get a little closer to the Coupé's real appeal. 'With low, sleek lines,' the copy read, 'it offers high power and individual luxury.' The car certainly did have sleeker lines than the Saloon, and these had a powerful appeal. It certainly did have plenty of power for its day, although of course the advertisement failed to mention that this power was offset by its great weight. And it certainly did offer individual luxury, with its extra instruments, rear seats strictly for two, the additional smoker's companion and the radio speaker in the

The optional badge bar, introduced in January 1961, is seen here on a 1965 3-litre Coupé, and proudly displays the badges of two enthusiasts' clubs.

rear parcels shelf. For all those features – and for the exclusivity which always accompanies the most expensive model of a car in its maker's range – buyers were prepared to pay a premium of slightly more than £150 (about 9 per cent of the list price before purchase tax) over the Saloon.

RIVALS

Those buyers who could afford a 3-litre Coupé but opted instead for something else were most likely to have been tempted away from Rover by the greater performance of a Jaguar. During the 1963 season, the obvious competition was from the 3.8-litre Mk II saloon, which offered prodigious performance and similar levels of interior trim at a saving of some £100 (about 7 per cent) over the Rover. From the autumn of 1963, the more obvious Jaguar rival was the new and rather less overtly sporting S-type, available with either 3.4-litre or 3.8-litre engine and in either case offering a lot more performance than the 3-litre Coupé could muster.

But probably not many buyers of a 3-litre Coupé were really interested in high performance. The car was fast *enough,* even though it was not particularly rapid. It was comfortable as a long-distance cruiser, it was prestigious and stylish, and it had a pedigree of respectability which still eluded Jaguar in the early sixties. The 3-litre Coupé was a *gentleman's* car, and its real competition came from cheaper offerings like the big Humbers, the Vanden Plas Princess, and Rover's own 3-litre Saloon and P4 models. The buyer who could afford the extra went for the extra style and exclusivity of the 3-litre Coupé – and it was those characteristics which won the model 5,419 sales over the three and a half seasons of its production.

PRODUCTION CHANGES

Not a lot changed in the specification of the 3-litre Coupé between its introduction in September 1962 and the end of production in December 1965. The minor running modifications are listed in the key production changes table in Chapter 4, and

3-litre Coupé: commission numbers and production figures

Note: The chassis number plates of these cars describe them as Mk II models. Nevertheless, they were always known simply as 3-litre Coupés until the Mk III models were introduced in 1965. There had never been a Mk I or Mk IA Coupé, so Rover saw no point in calling the car a Mk II until there was a risk of confusion with another variant. They are generally referred to today as Mk II Coupés.

1963 Coupé	735-00001 to -00237	Manual, home market	(237)
	736-00001 to -00015	Manual, RHD export	(15)
	738-00001 to -00058	Manual, LHD export	(58)
	740-00001 to -00180	Auto, home market	(180)
	741-00001 to -00018	Auto, export RHD	(18)
	743-00001 to -00038	Auto, export LHD	(38)
			Total: 546
1964 Coupé	735-00238 to -01188	Manual, home market	(951)
	736-00016 to -00044	Manual, RHD export	(29)
	738-00059 to -00219	Manual, LHD export	(161)
	740-00181 to -01430	Auto, home market	(1,250)
	741-00019 to -00110	Auto, export RHD	(92)
	743-00039 to -00121	Auto, export LHD	(83)
			Total: 2,503
1965 Coupé	735-01189 to -01910	Manual, home market	(722)
	736-00045 to -00064	Manual, RHD export	(20)
	738-00220 to -00304	Manual, LHD export	(85)
	740-01431 to -02628	Auto, home market	(1,198)
	741-00111 to -00189	Auto, export RHD	(79)
	743-00122 to -00162	Auto, export LHD	(41)
			Total: 2,145
1966 (Mk II) Coupé	735-01911 to -01956	Manual, home market	(46)
	736-00065 to -00067	Manual, RHD export	(3)
	738-00305 to -00312	Manual, LHD export	(8)
	740-02629 to -02770	Auto, home market	(142)
	741-00190 to -00206	Auto, export RHD	(17)
	743-00163 to -00171	Auto, export LHD	(9)
			Total: 225
			Total: 5,419

were broadly the same as for the contemporary Saloon. The most important difference was perhaps that there never was a 'b'-suffix version of the Coupé, and that all cars were 'a'-suffix types until the changeover to the 'c'-suffix specification occurred in March 1964. The reason was simply that the 'b'-suffix revisions in December 1962 affected only cars with manual steering; as all Coupés had power-assisted steering as standard, they quite logically retained their original designation.

THE 3-LITRE COUPÉ IN THE USA

The 3-litre Coupé was included in the 1963-season offerings by the Rover Motor Company of North America (RMCNA). Both overdrive and automatic versions were advertised, and although the cars did not differ substantially from other left-hand-drive export models, they did have a few special features. Like other cars destined for export, they had sealed headlamp units with prefocus bulbs in place of the sealed-beam units used on home market models. Broad-band whitewall tyres were certainly offered and were probably standard equipment, but the most noticeable special feature was the 'Rover' badge on the front wings ahead of the bright trim strip, exactly as on the Saloons shipped across the Atlantic at the time. Sales literature suggests that no two-tone colour schemes were offered, probably because the two-tone vogue of the fifties had by this time died in the USA.

The car was never a great sales success in the USA, although larger quantities were sold by RMCNA in Canada. RMCNA did try hard, however: at Easter 1963, the company prepared a special sales brochure for the Coupé in the shape of an Easter egg, to match the similar item for the Mk II Saloon. The 3-litre Coupé was withdrawn from the American market in 1964, when RMCNA switched its promotional efforts to the new 2000 saloon.

Just one of the major US motoring magazines reported in print on its road-test of a 3-litre Coupé. This was *Road and Track*, and the review which appeared in its October 1963 issue was mixed. The car went much better than the automatic Saloon model the magazine had tested in December 1959, reaching 105mph and taking 15.5 seconds to sprint from 0–60mph. But this was nothing special: the car only had 'enough performance to be acceptable though still not worthy of the term "sports sedan".' Build quality and refinement were very favourably seen, however: 'the Rover detailing, inside and outside, borders on the fantastic' and 'the engine is one of the smoothest and quietest in the world, better than any 6-cyl unit we can remember and virtually equal to the best of the V-8s.' Nevertheless, the testers preferred the Saloon to the Coupé, which in

> our frank opinion looks like a Californian Chop Job and would have come off much better if the 2.5in had been removed from the body section rather than the glass area. Headroom in the Coupé is minimal; two of our 6ft members proved that one's head tends to rub the headlining when sitting erect in either the front or rear seat.

THE MYTH OF THE 2.6-LITRE COUPÉ

Rover never did put a 2.6-litre Coupé into production, although it would have been perfectly possible to engineer one, using the engine from the P4 110 as was done for Saloon models. However, a story has arisen that some 2.6-litre Coupés were built for France. The origin of this story lies in a mistake in a sales leaflet issued by Rover's French importers, Franco-Britannic Autos of Levallois-Perret in the suburbs of Paris. The leaflet announces the new 2.6-litre Rover for the French market, and the small print makes quite clear that the engine will be available in the Saloon only and not in the Coupé; nevertheless, the illustration at the head of the page is of a Coupé!

It is interesting to note, too, that the Rover despatch books for the Mk II 2.6-litre Saloons record them as Coupés. Was there, perhaps, a plan for a 2.6-litre Coupé which never came to fruition but left its mark on the subconscious of people working for Rover at the time?

6 The Mk III Saloons and Coupés, 1965–1967

Despite the high regard in which the 3-litre was held, overall sales did drop during the 1965 model year. There can be little doubt that one reason was the changing face of the luxury-car market; the traditional wood-and-leather luxury barge was having to give ground to the new breed of more sporting 'executive' saloons, epitomized by Rover's own 2000 and the rival 2000 from Triumph. Perhaps most important for Rover, though, was the fact that the 2000's arrival had suddenly made the 3-litre look old. In the months which followed the 2000's October 1963 introduction and its enthusiastic acceptance, the 3-litre's styling, character

and performance all began to look as if they belonged to another era.

Rover already had plans to improve the P5's performance with the new V8 engine bought from General Motors during 1964, but there was a lot of work to be done and the company was aiming to have the car ready for the 1968 model year. In the meantime, sales of the six-cylinder Saloons and Coupés would have to be maintained at a reasonable level. The obvious answer was therefore to give the 3-litre a facelift for the 1966 model year; with any luck, this would keep it saleable for the two seasons before the new V8-powered version was ready.

This press release picture shows the Mk III Saloon (foreground) and Coupé together. Rover believed single-colour cars would predominate, and deleted two-tone paintwork from the catalogued options for the Saloon. Nevertheless, two-tone Coupés remained popular. The thicker rubbing strip no longer had a downturn ahead of the front wheel. The Saloon pictured has the standard rear seat, divided like the Coupé type into two individual armchairs; a bench was later introduced as an option to meet demand.

Rover 3-litre Mk III Saloon and Coupé (1965–1967)

Layout
Monocoque bodyshell with front subframe bolted in place. Four-seater or (saloon only) five-seater saloon, with front engine and rear wheel drive.

Engine

Type	3L7 with Weslake head
Block material	Cast iron
Head material	Aluminium alloy
Cylinders	Six, in line
Cooling	Water
Bore and stroke	77.8mm × 105mm
Capacity	2,995cc
Main bearings	Seven
Valves	Inlet valves in cylinder head and exhaust valves in cylinder block
Compression ratio	8.75:1
Carburettor	Single SU type HD6 (2in)
Max. power	134bhp gross (121bhp net) at 5,000rpm
Max. torque	169lb.ft at 1,750rpm

Transmission

Manual models	Hydraulically operated single dry plate clutch, 10in diameter
Automatic models	Torque converter

Internal gearbox ratios
Option 1 Four-speed manual with overdrive

	(Final drive 4.3:1)
Overdrive	0.77:1
Top	1.00:1
Third	1.27:1
Second	1.88:1
First	3.37:1
Reverse	2.96:1

Option 2 Three-speed automatic (Borg Warner type 35)

	(Final drive 3.54:1)
Top	1.00:1
Intermediate	1.45:1
First	2.39:1
Reverse	2.09:1

Suspension and steering

Front	Independent, with wishbones, laminated torsion bar springs and anti-roll bar
Rear	Semi-floating axle with progressive-rate semi-elliptic leaf springs

Steering	Burman recirculating ball type with variable ratio and power assistance
Tyres	6.70 × 15 crossply
Wheels	Five-stud disc type
Rim width	5in
Brakes	
Type	Discs at the front and drums at the rear, with servo assistance
Size	Disc diameter 10.75in
	Drum diameter 11in
Dimensions (in/mm)	
Track, front	55 (1,397)
Track, rear	56 (1,422)
Wheelbase	110.5 (2,807)
Overall length	186.5 (4,737)
Overall width	70 (1,778)
Overall height	59.25 (1,505) (Saloon)
	56.75 (1,441) (Coupé)
Unladen weight	3,738lb (1,695kg) (Saloon with overdrive)
	3,693lb (1,675kg) (Saloon, automatic)
	3,738lb (1,695kg) (Coupé with overdrive)
	3,702lb (1,679kg) (Coupé, automatic)

It is impossible to pinpoint when work started on the car which was to become the Mk III 3-litre. Some of the mechanical and cosmetic improvements it incorporated had been part of the 3-litre's rolling development programme and would have been introduced eventually in any case, while others were dreamed up simply to complete the package and to help draw attention to the new and improved models. However, it is undeniable that the major influence on the production specification of the Mk III 3-litre was the Rover 2000: both mechanical and cosmetic changes made for the MK III cars followed the lead of Rover's successful new high-volume saloon.

All car manufacturers try to save purchasing and manufacturing costs by using common components in as many different models as possible. At Rover, this approach had ensured that the Land Rover shared its petrol engine with the P4 saloons, that the P4 and P5 models shared their manual gearboxes and overdrives, and that smaller items such as brakes, door handles and switches were common to more than one model wherever possible. The P5 shared more components with the P4 saloons than with any other Rover or Land Rover model, and so when the P4 saloons went out of production in May 1964 it was inevitable that some changes to the P5 would follow. As the only other Rover saloon in production by this stage was the new P6 2000, it was equally inevitable that the 3-litre would soon start to share components with that.

To some extent, commonization of components between the P5 and P6 models was bound to be difficult: the P4 had been a car of similar size to the P5, but the P6 was an altogether smaller and lighter machine. Nevertheless, the Rover engineers set to with a will, and one of the areas of potential component sharing which they identified was the automatic transmission. The Borg Warner type DG in the 3-litre was in any case quite an old design by the mid-sixties, and the new type 35 was on trial for the forthcoming 2000 Automatic (which would appear in the autumn of 1966). Rover tried the type 35 behind the 3-litre engine, found it perfectly satisfactory provided a separate oil cooler was fitted ahead of the radiator, and decided to specify it for the Mk III 3-litre models which would be introduced the year before the 2000 Automatic.

Not the least of the type 35's advantages was that it could be used behind the higher-compression engine specified for the manual-transmission cars, and did not need the special low-compression type with its altered power and torque characteristics. This of course simplified production and saved costs, as Rover now needed to build fewer versions of the 3-litre engine. The type 35 transmission also needed less routine maintenance than the older DG, and Rover carefully engineered its installation to remove two disadvantages associated with the DG installation. So owners of 3-litres with the type 35 transmission found that they could check the transmission oil level and top it up from under the bonnet, instead of having to remove the transmission tunnel carpet and grovel about under the facia; and they also found that the transmission could be removed for major overhaul work without the need for the engine to be taken out first.

The introduction of the Borg Warner type 35 automatic gearbox gave the engineers an opportunity to make other transmission changes as well. While the manual cars with their overdrive gearboxes remained unaffected, a taller 3.54:1 final drive was fitted to the automatic 3-litres to give quieter and more economical high-speed cruising. The increased torque of the high-compression engine helped to maintain acceleration with this taller overall gearing, but Solihull made doubly sure that there would be no loss of performance by specifying for the type 35 lower intermediate gear ratios than those in the DG transmission. It all worked admirably well.

One other change introduced for the Mk III models was the result of overall Rover policy, and that was the change from a positive-earth electrical system to a negative-earth type. Both Rover saloons and Land Rovers went over to the new system,

---

Optional extras available for the Mk III 3-litre models

Badge bar
Bench rear seat (from June 1966)
Dipping interior mirror
Electric immersion heater for cylinder block
Exhaust tail pipe finisher
Extension speaker for radio (fitted under rear parcels shelf), with balance control
Floor mats, beige nylon fur
Floor mats, charcoal grey rubber
Fog lamp
Foot pump
Head restraints, front (Saloon and Coupé) and rear (Saloon only) (from spring 1966)
Heated rear window
Laminated windscreen
Mud flaps, front and rear
Radio (Pye or Radiomobile)
Seat belts (Irvin) for front and rear
Special paint finish or interior trim to customer's requirements
Spot light
Towbar
Two-tone paint
Whitewall tyres
Wing mirrors

which was introduced to improve compatibility with the latest transistorized electrical components, although Rover actually made the change in stages. The Mk III 3-litres were the first of Solihull's products to take on the new system when they were introduced in autumn 1965; the P6 saloons changed early in 1966 and Land Rovers went over to negative-earth systems about a year after that. To prevent servicing errors, the 3-litres were provided with large stickers reminding of their negative earth electrical systems. They also took on a heavily revised wiring harness with a new fuse box containing no fewer than twelve fuses instead of the previous two. (The early fuse box had commonly been supplemented by a pair of separate fuse holders to cover the overdrive and radio circuits.).

One particular area where the 2000 had made the 3-litre look old-fashioned was in its interior. The 3-litre's seats were broadly similar to those of the P4 – in fact the final P4s had shared their individual front seats with the Mk II 3-litres – and during 1964 or 1965, David Bache had consulted the Mulliner Park Ward coachbuilding division of Rolls-Royce for some ideas about updating the P5's interior. Some sketches were prepared, but it must have already been obvious at Solihull that a new interior which copied the best features of the 2000 would be a much cheaper option. So the Mulliner Park Ward suggestions were sidelined (and the actual sketches have disappeared), and Bache decided to redesign the interior of the P5 to give it a family resemblance to the much-liked 2000 interior.

The main change he made was to the seats. The 2000 had been designed as an uncompromising four-seater, with individual front seats and individually-shaped rears, and what Bache did in essence was to scale the 2000's front and rear seats up to suit the bigger P5. In this process, the Saloons lost both their bench front seat option and their standard bench rear seat (although a bench rear seat option had to be introduced later to meet customer demand), and both Saloon and Coupé models took on four sumptuous, supportive armchairs. The front seats had the Bache-patented backrest friction lock which had been so highly praised in the 2000 because it offered infinite adjustment instead of the traditional stepped adjustment, and they also had a winding handle which could raise or lower the cushion. Careful shaping of their backrests had also freed

Rover liked to issue artist's drawings as press material in the early sixties, and many showroom catalogues were composed of drawings rather than actual pictures. These are the press release pictures of the new Mk III facia, with the clock now in the corner fillet (left) and of the Mk III seats (right), which were derived from those originally designed for the Rover 2000.

a very worthwhile 2½in (6.35cm) of extra legroom for rear seat passengers as compared to the earlier cars, which made its own very obvious contribution to the more luxurious feel of the P5's interior. These backrests also incorporated fittings to receive head restraints, although these would not become available until some months after production had started. Always covered in leathercloth rather than leather, these head restraints also incorporated a reading lamp for the rear seat passenger.

Increased luxury was very much on Bache's mind when he developed the Mk III interior, and he incorporated a number of small but very pleasing touches which remain a delight to owners more than 30 years after their introduction. While Coupés retained the 'smoker's companion' of lidded ashtray and lighter between the rear seats, the top of the central bolster cushion on Saloons was hinged so that it lifted to reveal a useful small storage tray. Behind the rear-seat armrest on both Saloons and Coupés, Bache added a fold-out picnic tray with an extending cup-holder; he matched this by fitting a veneered lid to the under-dash tool

tray and relocating it in the middle, where it could serve as a picnic tray for the front seat occupants. Both Saloons and Coupés also gained a rear compartment heater, plumbed in to the cooling system like the main heater. Fitted in the seat pan and blowing hot air through a pair of outlets in the rear heelboard, it was under the control of the rear seat occupants from a rotary switch located on the transmission tunnel just behind the front seats. The console containing that switch was also fitted with a blanking plug, and removing that exposed the mounting for a volume control which could be fitted in conjunction with the optional rear extension speaker for the radio. The rear parcels shelf – now slightly recessed and covered with leathercloth on both Saloons and Coupés – was already pre-punched with a network of holes to allow sound to come from the speaker which could be fitted beneath its centre.

Only one thing was lost in this radical rearrangement of the interior trim, and that was the map-pocket which had been fitted to the back of each front seat on earlier cars. David Bache's restyle had incorporated plain panels behind the seats, and even the

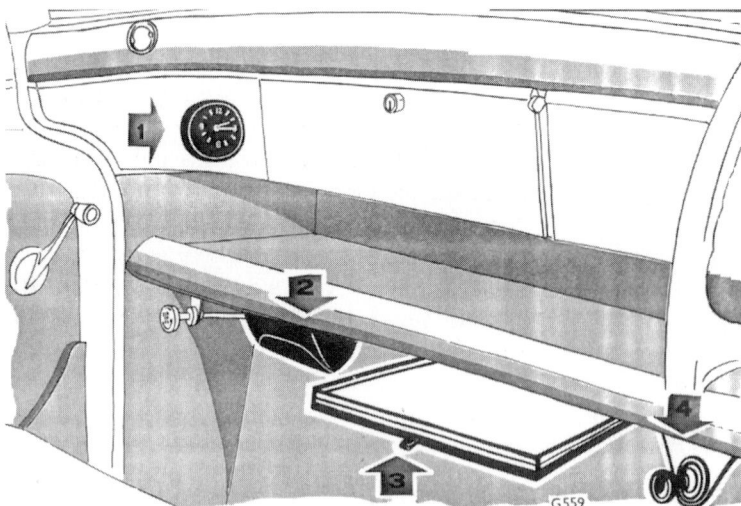

This picture from the Mk III Owner's Instruction Manual shows two of the model's new features. Number 1 is the clock, now recessed into the corner wood fillet instead of perched on the centre of the crash rail, and number 3 is the tool tray, now repositioned centrally and featuring a veneered lid so that it could double as a picnic tray. Numbers 2 and 4 are of course the ashtray and cigarette lighter respectively.

This Rover publicity picture says as much about the kind of customer the 3-litre was aimed at as it does about the car! The three 'pips' on the rear wing are clearly visible here, and a close look shows that this Saloon was fitted with a heated rear window: on these early laminated types, a thin silver line ran across the vertical heating elements, about a third of the way up the window.

ashtrays had been banished to the rear door trims, where a slight restyle had been necessary to incorporate them. The facia remained basically untouched, although even here Bache had made a few changes. The clock – now a new Smith's type incorporating its own battery and self-charging mechanism – had moved from the centre of the crash-rail to the wooden fillet in the corner of the dash where it was less likely to cause injuries in an accident. A stainless steel reinforcing edge had been added to the roll at the edge of the front parcels shelf, and finally the oil-level check switch had been deleted from the instrument panel, to be replaced by the reserve petrol pump changeover switch. New petrol gauges on both Saloon and Coupé of course no longer incorporated the yellow marking for the oil-level indicator.

All this had to be wrapped up in an exterior package which would make the Mk III cars stand out. It looks as if the budget which David Bache was granted to achieve this was rather limited, or perhaps it was simply that the interior makeover had used up the lion's share of the Styling Department's budget for the Mk III models. One way or another, the external changes sanctioned for production were minimal – although it says a lot for Bache's skill as a stylist that they made the

Mk III cars instantly distinguishable from their predecessors.

The most obvious external change was to the side trim strip. On earlier P5s, it had always been a thin strip of bright metal, terminating just ahead of the front wheelarch with an angled downward extension which made an ideal colour boundary for two-toning of the lower body. However, two-tone paintwork of the type popular in the fifties was now beginning to look dated, and the trend in the sixties was towards single exterior colours. So Bache fitted a wider trim strip for the Mk III models, which broke up the large flat expanse of the P5's lower body sides but had no downward extension at the front to act as a colour boundary. A neat and mildly amusing touch at the back of the car replaced the upturn of the early trim strip with three pips, which signified the P5's third incarnation. And in case anyone was in any doubt, there were 'Mark III' badges on the front wings and bootlid to identify the car. As it was impossible to two-tone the lower body sides satisfactorily with this new trim strip, Rover offered two-tone Saloons with the contrasting colour covering the roof only, as it always had on Coupés. In practice, however, two-toned Mk III Saloons did not look very attractive. Few were

117

Paint and trim colours, Rover 3-litre Mk III

1966 model year

There were ten standard paint colours and six standard interior colours with carpets to match. Both Saloon and Coupé models were available with ten two-tone paintwork combinations. The standard combinations for single-tone cars were:

Body colour	*Upholstery*
Black	Blue, Green, Red, Stone or Tan
Burgundy	Grey, Red or Stone
Charcoal	Blue, Green, Grey, Red or Stone
Juniper Green	Green, Grey, Stone or Tan
Light Navy	Blue, Grey or Stone
Marine Grey	Blue, Green, Grey, Red, Stone or Tan
Pine Green	Green, Grey, Stone or Tan
Stone Grey	Blue, Green, Red or Tan
Steel Blue	Blue, Grey, Red or Tan
White	Blue, Green, Red, Stone or Tan

The combinations for two-tone Saloons were as follows:

Lower body	*Upper body*	*Upholstery*
Charcoal	Marine Grey	Blue, Green, Grey, Red, Stone or Tan
Juniper Green	Pine Green	Green, Grey, Stone or Tan
Light Navy	Steel Blue	Blue, Grey or Stone
Marine Grey	Black	Blue, Green, Grey, Red, Stone or Tan
Marine Grey	Light Navy	Blue, Grey or Stone
Marine Grey	White	Blue, Green, Red, Stone or Tan
Pine Green	Marine Grey	Green, Grey, Stone or Tan
Steel Blue	Charcoal	Blue, Grey or Red
Stone Grey	Burgundy	Grey, Red or Stone
Stone Grey	Juniper Green	Green, Grey, Stone or Tan

The combinations for two-tone Coupés were as follows:

Roof	*Lower body*	*Upholstery*
Charcoal	Black	Blue, Green, Grey, Red, Stone or Tan
Charcoal	Steel Blue	Blue, Grey or Red
Juniper Green	Pine Green	Green, Grey, Stone or Tan
Light Navy	Marine Grey	Blue, Grey, Stone or Tan
Marine Grey	Black	Blue, Green, Grey, Red, Stone or Tan
Pine Green	Stone Grey	Green, Stone or Tan
Steel Blue	Light Navy	Blue, Grey or Stone
Steel Blue	White	Blue, Grey, Red, Stone or Tan
Stone Grey	Burgundy	Grey, Red or Stone
Stone Grey	Juniper Green	Green, Grey, Stone or Tan

Paint and trim colours, Rover 3-litre Mk III continued

1967 model year
There were seven standard paint colours and four standard interior colours with carpets to match. Saloon models were always single-tone unless to special order; Coupés were available in all seven single-tone colours and in seven two-tone paintwork combinations. The standard combinations for single-tone cars were:

Body colour	Upholstery
Admiralty Blue	Buckskin, Buffalo, Sandalwood or Toledo Red
Arden Green	Buckskin, Buffalo or Sandalwood
Bordeaux Red	Buckskin, Buffalo, Sandalwood or Toledo Red
Burnt Grey	Buckskin, Buffalo, Sandalwood or Toledo Red
Juniper Green	Buckskin, Buffalo or Sandalwood
Silver Birch	Buffalo, Sandalwood or Toledo Red
White	Buffalo, Sandalwood or Toledo Red

The combinations for two-tone Coupés were as follows:

Roof	Lower body	Upholstery
Admiralty Blue	Burnt Grey	Buckskin, Buffalo, Sandalwood or Toledo Red
Burnt Grey	Silver Birch	Buffalo, Sandalwood or Toledo Red
Burnt Grey	White	Buffalo, Sandalwood or Toledo Red
Juniper Green	Arden Green	Buckskin, Buffalo or Sandalwood
Silver Birch	Admiralty Blue	Buckskin, Buffalo, Sandalwood or Toledo Red
Silver Birch	Bordeaux Red	Buckskin, Buffalo, Sandalwood or Toledo Red
Silver Birch	Juniper Green	Buckskin, Buffalo or Sandalwood

According to the 1967-season sales catalogue, 'the threepenny piece shown in the picture is not a standard feature but its ability to stand upright with the engine running, is'. On the Mk III models, the windscreen washer reservoir had moved to the opposite side of the engine bay (the far side in this picture) and there was a tap to provide hot water for the rear compartment heater. It can just be seen between the rocker cover breather and the wiper motor on the right of this picture.

ordered, and two-toning was deleted as an option for Saloons after a year.

Bache made just two more changes to the exterior of the Mk III models. On the grille, he replaced the original central strut and badge with a broader strut containing the 2000's grille badge. And at the back, he modified the rear scuttle panel to accept a chromed, circular filler cap with a press-button release. This was not identical to the 2000's filler, but it was broadly similar and helped to reinforce the family identity between the two cars. As the filler needed modification anyway – it had always been prone to leaks on the later Mk II models – Bache no doubt seized his opportunity to make one more cosmetic change on the back of a necessity.

(Above) The centre spine of the Mk III grille broadened out to accommodate a Rover 2000-type Viking ship badge, and the grille's top bar no longer carried any badging.

(Right) This 'Mark III' badge was found on the front wings and on the bootlid.

THE MK III MODELS ON SALE AND ON TEST

Despite Rover's best efforts, the sales slide which had started in the 1965 season continued. The revitalized Mk III models probably did little more than prevent the slide becoming a disaster – but if that was the case, then they certainly fulfilled their role in bridging the gap until the V8-engined models were ready. The 1965-season figures had shown a drop of around 15 per cent in overall 3-litre sales as compared to the previous season, and the 1966 season showed a further drop of 30 per cent in Saloon sales and 10 per cent in Coupé sales. The 1966-season sales figures included the final Mk II models which remained on sale until the end of the 1965 calendar year, so it was abundantly clear that the Mk IIIs had not reversed the downward trend. For the 1967 season, the Mk IIIs did even worse: Saloon sales were down by over 50 per cent during 1966, while Coupé figures were down by nearly 60 per cent. No doubt rumours of the forthcoming V8-engined models helped keep sales depressed during the early months of 1967, but the picture must have been a depressing one for Rover. It was touch and go whether sales really would hold up until the re-engined models came out in mid-1967.

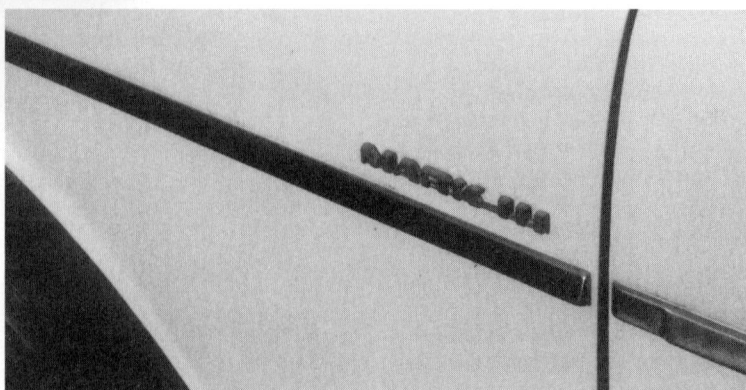

Britain's motoring magazines tended to pull their punches in road tests in the mid-sixties, but it was usually possible to work out what were the worst features of the cars they tested by reading between the lines of what appeared in print. Yet there is almost nothing in contemporary road tests of the Mk III 3-litre which would suggest why the car did not sell well. *Autocar* summarized its impressions of an overdrive Coupé tested in the issue of 6 May 1966 by noting that the car rolled too much on corners and that the brakes had not stood up well to a newly-introduced fade test, but it also noted that this was an 'exceptionally quiet car offering armchair comfort for four' and that it was 'a fast car providing really relaxed travel.' *Motor* was equally complimentary about the car's virtues in its test of an automatic Saloon dated 19 March 1966, noting that 'although the 3-litre is now eight years old, it is still one of the most refined saloons in the world for comfort and quietness' and that its new interior appointments had 'raised the

standard of comfort on the Mk III to such a high level that you can forgive the slightly turbulent ride on bad roads – one of the few things that betray the design's age.'

Set against these comments, however, was the car's image. *Motor Sport* summed it up neatly in its October 1966 issue, headlining its impressions of an overdrive Coupé, 'For older men.' Road tester Bill Boddy was in any case a trifle miffed at having to accept a 3-litre for test when he had asked for a 2000TC (which an accident had made unavailable), but his overall impression was that 'this big Rover seems to me to be an inexpensive luxury car for the older motorist'. The *Motor* test had hinted at the same thing, suggesting that the 3-litre was an elderly gentleman's car and arguing that, despite its impressive performance, 'it is a car for relaxation, not exhilaration.' In the thrusting new age of the sixties, the 3-litre had come to symbolize the past rather than the future – and yet it was still highly regarded by those whose priorities lay with comfort rather than with speed. As Bill

Mk III Saloon: commission numbers and production figures

1966 Mk III Saloon	795-00001 to -00576	Manual, home market	(576)
	796-00001 to -00038	Manual, export RHD	(38)
	798-00001 to -00098	Manual, export LHD	(98)
	800-00001 to -01558	Auto, home market	(1,558)
	801-00001 to -00101	Auto, export RHD	(101)
	803-00001 to -00136	Auto, export LHD	(136)
			Total: 2,507
1967 Mk III Saloon	795-00577 to -00823	Manual, home market	(247)
	796-00039 to -00088	Manual, export RHD	(50)
	798-00099 to -00159	Manual, export LHD	(61)
	800-01559 to -02388	Auto, home market	(830)
	801-00102 to -00245	Auto, export RHD	(144)
	803-00137 to -00216	Auto, export LHD	(80)
			Total: 1,412
			Total: 3,919

Mk III Coupé: commission numbers and production figures			
1966 Mk III Coupé	805-00001 to -00425	Manual, home market	(425)
	806-00001 to -00007	Manual, export RHD	(7)
	808-00001 to -00065	Manual, export LHD	(65)
	810-00001 to -01117	Auto, home market	(1,117)
	811-00001 to -00048	Auto, export RHD	(48)
	813-00001 to -00044	Auto, export LHD	(44)
			Total: 1,706
1967 Mk III Coupé	805-00426 to -00557	Manual, home market	(132)
	806-00008 to -00013	Manual, export RHD	(6)
	808-00066 to -00087	Manual, export LHD	(22)
	810-01118 to -01637	Auto, home market	(520)
	811-00049 to -00129	Auto, export RHD	(81)
	813-00045 to -00078	Auto, export LHD	(34)
			Total: 795
			Total: 2,501

Boddy admitted, 'several of my friends whom I regard as discerning drivers love their 3-litre Rovers, or profess to'.

RIVALS

Prices of the 3-litre Mk III models in Britain remained constant throughout the production run, with the cheapest manual Saloon priced at £1,520 before purchase tax and the most expensive automatic Coupé costing £1,651 10s 0d without extras and before tax. Direct rivals on price were the S-type Jaguars, the 3.4-litre overdrive model costing £1,468 and the 3.8-litre automatic £1,608, again before tax. Both were very much quicker cars than the 3-litres, and were considerable bargains although their styling was not universally liked.

More directly comparable in terms of performance and appointments were the Vanden Plas 4-litre R from BMC, carefully priced at £1,650 before tax, and the big Humbers. In autumn 1965, a Super Snipe

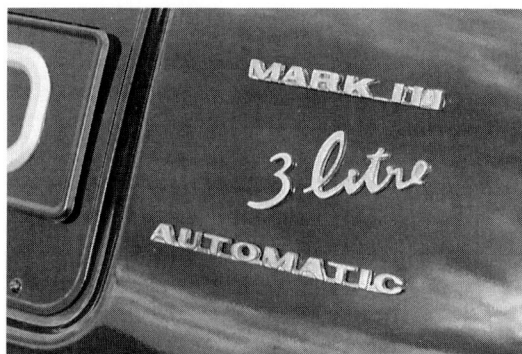

It is at least arguable that the bootlid badging on automatics had become rather fussy. All models carried the 'Mark III' badge above the 'Coupé' or '3 Litre' script, and Borg Warner-equipped cars had an additional identifier below that.

limousine cost £1,425, an Imperial saloon £1,565 and an Imperial limousine £1,690. However, the Vanden Plas 4-litre R never lived up to the promise of its Rolls-Royce engine, and production had dropped to 200 cars a year by 1967. The big Humbers also sold so poorly that their prices had to be slashed dramatically: by the time of the Earls Court Motor Show in October 1966, the most expensive Imperial limousine was down to £1,547 and the cheaper models had been repriced accordingly. Quite clearly, these were bad times for the traditional British luxury car, and against this background the declining sales of the 3-litre Rover are much easier to understand.

PRODUCTION CHANGES

Relatively few specification changes were made during the production of the Mk III models, not least because all P5 development effort at Solihull was now focused on the

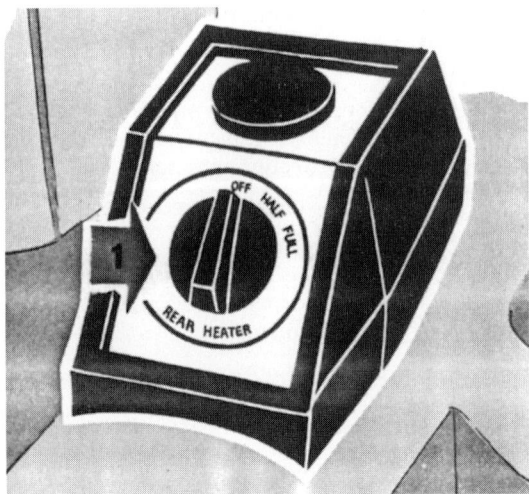

Rear passengers in the Mk III had their own heater blower control, sited on the transmission tunnel between the front seats. The heater itself was under the rear seat and was switched on or off by a tap at the rear of the engine bay. This picture is taken from the Owner's Instruction Manual. *The black circle on top of the control box was actually a decorative blanking plug; it could be removed and replaced by a rotary volume control for the rear radio speaker.*

forthcoming V8-engined models. The full list is contained in the table, but it is clear that the major changes introduced in the first twelve months of production were to meet customer demand: lower front seat cushions increased the headroom available and the bench seat option reinstated the five-passenger capacity of the Saloon models. For the 1967 season, the changes were designed to streamline production and reduce costs: there were fewer paint and trim options, and all the new colours were drawn from the range available for the P6 models.

THE MK III IN THE USA

By the time the Mk III 3-litres became available, the focus of Rover activity on the other side of the Atlantic had switched firmly to the 2000. The new car had been rapturously received by the motoring press (although customers were less easy to find), and the Rover Motor Company of North America knew by this stage that no amount of promotional activity was going to turn the 3-litre into the sort of best-seller that the 2000 might turn out to be. So, although the Mk III Saloons and Coupés did go on sale in the USA and in Canada, they found few takers.

The NADA (North American Dollar Area) Mk IIIs had few differences from cars for the rest of the world. Like the final Mk II models, they came with narrow-band white-wall tyres, a dipping rear-view mirror, a vitreous-enamelled exhaust manifold, laminated windscreen glass and prefocus-bulb headlamps in sealed lens-reflector units as standard. Sales catalogues note that tinted glass was standard, although contemporary parts catalogues do not confirm this. Air conditioning – fitted by the importers – was also available as an optional extra, and the optional radio was a Blaupunkt AM-FM Custom set.

123

Rover 3-litre Mk III Saloon and Coupé: production changes

Note: All dates given for changes where no chassis number is quoted must be treated as approximate. The dates are taken from issues of *Rover Service Newsletter*, which usually reported changes between one and four months after they had actually been made on the assembly lines. Dates given for changes where a chassis number is quoted are exact for home market models and reflect the date into Despatch (i.e. the date the car was transferred from the assembly lines to the Despatch Department) given in records held by the British Motor Industry Heritage Trust.

Date	Home, Manual	Home, Auto	Other	Remarks
September 1965	Mk III Saloon and Coupé models introduced.			
				Borg Warner type 35 automatic transmission replaced DG type (serial number prefix 3EU, on white plate).
January 1966				¼in UNF retaining screws replaced 10 UNF screws in front seat adjusting block, to prevent slippage of tie rods.
February 1966	805-00225A	810-00474A	806-00004A 808-00018A 811-00015A 813-00013A	Modified front door locks on Coupés, with ball bearings and springs deleted to prevent seizure.
				Modified flexible pipes on automatic transmission oil cooler to eliminate noise.
May 1966				Front- and rear-seat headrests available.
June 1966				Larger-section throttle lever for automatic transmission cars, to prevent distortion.
				Improved anti-squeal shims and chromium-plated pistons for front brake callipers.
				Additional negative earth warning labels fitted alongside battery and fuse box.
				Bench rear seat introduced as an option for Saloons.
				Redesigned tool tray, without small centre lever at front; wood-finish top removable only when tray has been removed.

1967 model-year

August 1966				Automatic transmissions modified to improve change quality and gear selector feel, from transmission number 3EO3398.
				Lower front-seat cushions available (identified by the absence of a vertical zip on the front face).
				New body and trim colours; two-tone Saloons no longer available.
				Paint touch-up pencils discontinued for new cars.
December 1966				Softer rear engine mounting rubbers introduced for automatic models only; retro-fit possible on Mk III models.
				Differential housing locating dowels deleted from rear axle casing.
March 1967	805-00506A	810-01502A	808-00074A 811-00109A 813-00063A	Redesigned lower BC post trim on Coupé models to accommodate safety harness anchorage points.
May 1967				Improved windscreen wiper speed switch; retro-fit possible for Mk IIc and Mk III.
July 1967				Improved windscreen wiper on/off switch; retro-fit possible for Mk IIc and Mk III.
August 1967				Improved seals on power steering box, from steering box number 46321.
January 1968				Brass-bodied water temperature transmitter introduced as service replacement for steel and aluminium type, to prevent corrosion.
March 1968				Sandwich-construction rocker cover gasket introduced as service replacement to facilitate assembly and prevent leaks.

7 Development of the P5B

The way Bruce McWilliams tells the story, Rover's acquisition of the General Motors light-alloy V8 engine which replaced the 3-litre straight-six in the P5 models during 1967 was, partly at least, down to his insistence that Rovers needed more power. McWilliams had been appointed to run the Rover Motor Company of North America in 1963, and as part of his duties he reported back to Solihull with ideas about how best to improve sales of Rovers and Land Rovers in the USA and Canada.

One of the biggest problems which the Rover products of the early sixties faced was a lack of performance. In the USA, straight-line performance was becoming a very important factor in selling cars, and the 'muscle-car' craze for enormously powerful cars was just beginning. As the road tests quoted in earlier chapters make clear, US motoring journalists were always rather underwhelmed by the performance of the 3-litre. The P4 saloons which the company also attempted to sell in the USA up to 1964 were even slower, and the new P6 2000 which replaced them may have had excellent handling and more than respectable performance, but it was a hard job for RMCNA to promote its puny 2-litre four-cylinder engine as an alternative to the 6-litre and 7-litre V8s which Detroit was turning out at lower prices. As for the Land Rover, its road performance was simply pathetic; American owners who wanted to use one during a weekend vacation would often hitch the

Land Rover to an A-frame behind the family car and tow it to their destination in order to get there more quickly and in comfort!

McWilliams lost no time in putting his thoughts across to William Martin-Hurst, who was by then Rover's Managing Director. Martin-Hurst took a rather broader vision of Rover's products than his predecessors had done, and he listened to what McWilliams had to say. 'We had initially attempted to get Rover to buy a Chrysler V8 engine to give their vehicles more power and smoothness,' remembered McWilliams in 1994. That plan fell through, but Martin-Hurst remained receptive to the idea of putting an American V8 engine into Rovers.

The next stage happened very soon after that. Beginning in 1962, Rover had been selling small numbers of Land Rover diesel engines to the Ruston company in Lincolnshire for conversion to marine use in small boats, and Ruston's branch in Toronto showed one of these engines to Carl Kiekhaefer, head of Mercury Marine in the USA. Kiekhaefer decided to negotiate a contract for similar engines to be marketed by his own company, and it was in connection with these negotiations that William Martin-Hurst flew out to the USA during 1964. As he used to tell the story himself, he was visiting the Mercury Marine premises at Fond du Lac, Wisconsin, when he spotted a small-block V8 engine on the floor of the workshop. On asking what it was, he was told it was a General Motors 215 cubic-inch V8 which had been taken out of a Buick Skylark; Mercury Marine were planning to look into the possibilities of adapting it for marine use because the engine had just gone out of production and General Motors might be interested in selling the manufacturing rights.

Martin-Hurst measured the engine for size, and very quickly recognized that it was small enough to fit into the engine bays of both the P5 3-litre and the P6 2000. So he asked Kiekhaefer if he could take it away with him, had it crated up and arranged for it to be shipped back to Solihull. However, he soon found that his colleagues back in the UK did not share his enthusiasm. Market analyst Graham Bannock, who was then doing the initial market research for the vehicle which would become the Range Rover in 1970, remembers that the Rover people reacted with horror. 'What? Use a foreign engine in a Rover? A non-Rover engine, let alone an American engine!'

Technical Director Peter Wilks also insisted that his engineers were far too busy to do anything with the General Motors V8. They were heavily involved with looking at no fewer than three derivatives of the 2000's overhead-camshaft four-cylinder engine: a twin-carburettor version, a 3-litre, six-cylinder derivative, and a five-cylinder 2½-litre as well. While Martin-Hurst had to accept that, he was not to be put off so easily. So he had a word with Ralph Nash, Superintendent of the Engineering Workshops, and persuaded him – with Peter Wilks' grudging agreement – to fit the Buick engine into a car. Nash's staff squeezed the V8 into a 2000 development hack during November 1964, and quite quickly had the vehicle up and running.

More and more persuaded that the Buick V8 offered precisely what Rover needed, Martin-Hurst now set about persuading his fellow Directors of its worth. One anecdote often retold is that he drove the V8-powered 2000 down to London for a Board meeting at

Devonshire House in Piccadilly one day, and then asked Rover's Chairman, Spencer Wilks, if he would like to drive it back to the Midlands. Wilks was astounded by the car's performance, and is alleged to have told Martin-Hurst that it was the very first Rover he had ever driven which was not underpowered!

By stealth, Martin-Hurst gradually won his colleagues on the Rover board round to his way of thinking, and the results from the test programme which he managed to get going for the V8-engined car during 1965 amply confirmed his faith in the engine. However, things were still not plain sailing. Martin-Hurst attempted to open negotiations with Phillip Copelin at GM, and found he was getting nowhere fast. Jim Dodsworth, who had been Project Engineer on the 3-litre Coupé but had by then moved on, remembers Martin-Hurst telling him that he eventually decided to call on Copelin, and patiently sat in the waiting-room outside the GM executive's office for some considerable time until the American could find a few minutes to see him. It transpired that the negotiations had failed to get off the ground because GM simply did not believe Rover could be serious in wanting to negotiate for the manufacturing rights to one of their redundant engines.

From here on, however, things happened rather more quickly. GM were only too pleased to make some money out of this engine they no longer wanted, and a deal was signed in January 1965. The Rover engineers were immediately given access to all the engineering drawings and service records for the engine, and a quantity of brand-new engines (there were allegedly 39) were taken out of store and shipped across to Solihull to form the basis of a comprehensive development programme. Some of the original production-line tooling was also transferred to Rover. And finally, Martin-Hurst negotiated with GM to bring to Britain on a short-term contract the man who knew more about the light-alloy V8 than anyone else – Joe Turley, Head of Engine Design at Buick.

Turley was then about eighteen months from retirement anyway, and was somewhat bemused to find himself installed at Solihull as a consultant to the Rover engineers who were busily adapting the V8 engine to suit its new home. His advice and knowledge proved invaluable, although he took some convincing at first that Rover needed to improve the engine's breathing to get a higher rev limit than the original 4,400rpm of the American engine; in his experience, people simply did not drive that fast. Rover's Chief Engine Designer Jack Swaine remembers that Turley was finally convinced after test driver Philip Wilson took him for a 100mph-plus run up the new M6 motorway in a six-cylinder P5. Turley was then 'prepared to concede that maybe the engine did need to do more than 4,400rpm, and what's more it had got to live at that,' Swaine told journalist Martin Buckley in 1988.

THE BUICK V8

The engine which Rover knew as the '2158' (215 cubic inches, 8 cylinders) had seen only three years' production in General Motors cars and was therefore a relatively young design. It was also an advanced design, with the extra advantage of remarkable simplicity and light weight on account of its all-alloy construction. In its issue of 23 September 1960, *The Autocar* had enthused about the then-new engine, prophesying that it would be the most widely copied engine in the next ten years.

The small-block V8 had been developed in the late fifties by General Motors' Buick division, although some of the principles of its construction dated back to the beginning of the decade and others almost certainly derived from the BMW alloy V8 introduced in 1954. It had been designed to power the so-called 'compact' saloons which appeared in the early sixties as Detroit's response to the small and nimble European cars which had then begun to make such an impact in the USA; for this reason it was smaller than the average American V8 and was made of aluminium alloy to save weight. However, two things had conspired to bring about its early demise. The first was that Detroit had developed new thin-wall casting techniques which reduced the weight of iron-block engines to not much more than the aluminium V8; as cast-iron engines were much

The Buick engine was found in a number of General Motors' 'compact' models of the early sixties. This is a 1961-model Buick Special which, with its 112in wheelbase, was close in size to the big Rover.

The first development cars to use the Buick engine at Rover were converted 2000s. This one – with an experimental headlamp arrangement – was used for long-distance tests in continental Europe during 1965. Note the large pancake air cleaner of the Buick engine, which presumably still had its original Rochester carburettor. The blocks of foam on the front panel and radiator were a quick way of getting around installation problems!

cheaper to make than alloy types, there was no question about which way to go. The second was that the American public's flirtation with 'compact' saloons had proved short-lived, and the small-block V8 simply did not fit the demands of a market which was rapidly becoming power-hungry. So the 'B-O-P' (Buick-Oldsmobile-Pontiac) 215-cubic inch engine ceased production in 1963, after being fitted to some 750,000 Buick Specials (of which the Skylark was a derivative), Oldsmobile F-85s and Pontiac Tempests.

Even though the engine's design was fundamentally simple, it took the Rover engineers some time to get to grips with the characteristics of an all-alloy 90-degree V8 with overhead valves operated by a single central camshaft. Their experience in the early sixties was limited to in-line engines made of cast-iron, although some had aluminium alloy cylinder heads. There were all kinds of problems to be resolved, as well. The Buick engine was manufactured by American methods, the blocks being gravity die-cast with their cylinder liners held in place. British manufacturing technology was different, and

the engine had to be adapted so that it could be sand-cast and have press-fit liners fitted afterwards. While adapting the block for this different method of manufacture, the Rover engineers also allowed for the increased stresses of more powerful or larger-capacity variants of the V8 by adding extra metal around the main bearings.

British ancillaries had to replace the American ones, too, and Lucas Industries were approached to provide ignition components to replace the American-made AC-Delco originals. Then there was the question of carburation. Test Engineer Brian Terry remembers that early P6 development cars fitted with Buick engines used to die on one cylinder bank when they were cornered hard because the single Rochester carburettor was unable to deliver fuel under enough pressure to counteract the roll of the body! Twin British-made SU carburettors proved the natural choice, and a pent-roof manifold was drawn up to accommodate them. The two men most closely involved with the V8 in its early days at Solihull, Dave Wall and Richard Twist, happily

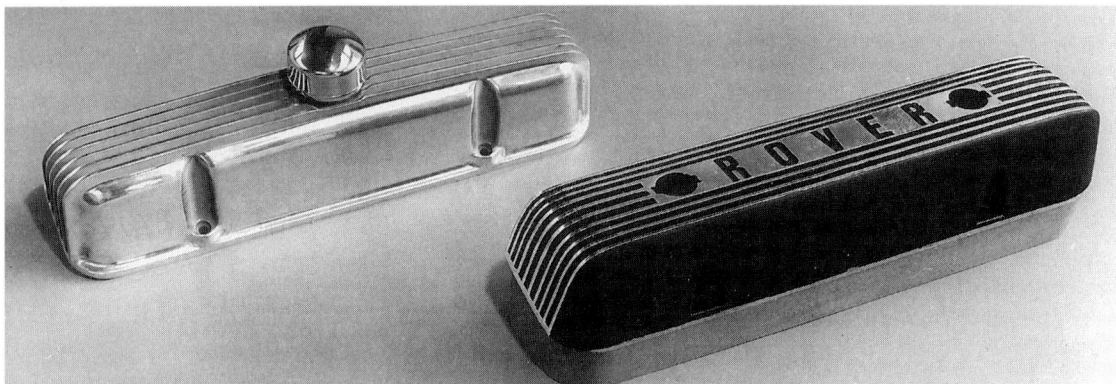

While the Rover engineers adapted the American V8 for English manufacturing methods, the stylists got down to their side of the business. Here is a proposed rocker cover design alongside an American original. It was a slightly different design which was actually adopted for production.

admit that they borrowed the design of this from the contemporary Rolls-Royce V8 engine. The self-adjusting hydraulic tap-

pets were unlike anything made in Britain at the time, and so Rover settled for import-ing them (together with camshaft blanks,

This is the Rover V8 engine as it went into production for the P5B. The bent-over blades of the cooling fan were designed to cut down noise, but the fan was redesigned with straight blades later on. The Borg Warner auto-matic transmission is bolted to this engine, and its filler tube and dipstick can be seen behind the exhaust manifold. The pipes leading from the transmission and terminating near the oil filter are those for the gearbox oil cooler mounted in the radiator. The choke cable has been removed to make the engine look tidier; normally it came across the engine from the driver's side and met a junction box clipped to the inlet trumpet of the air cleaner, from where short cables went to each of the twin SU carburettors.

the timing gears and timing chain) from the Diesel Equipment Company of Grand Rapids, Michigan.

The Rover V8 did not finally enter production at the company's Acocks Green engine plant until the early months of 1967, but the engines supplied by Buick were put to good use in the meantime. The majority were 'Roverized' as soon as possible, with components which Solihull intended to use in the production engine, and an intensive test programme was carried out between 1965 and 1967 using V8-powered P6s, P5s and Land Rovers.

The first Buick V8 to reach Solihull had gone into a P6 development car, mainly because Martin-Hurst hoped to use it to give the company's latest saloon a strong appeal to American customers. However, it was the ageing P5 which more urgently needed a shot in the arm, even though it was unlikely ever to become a big seller in the USA. So V8 engines very quickly found their way into some P5 development cars. One of them was a well-worn Experimental Department hack

registered 165 ENX, which was given engine number 2158/17 (215 cubic inches, 8 cylinders, engine number 17) and then scrapped on 23 December 1966. Brian Terry vividly remembers testing this or another car like it, sitting in the back taking notes from a bank of test gauges while someone else drove as fast as possible up the M6 motorway from the Midlands to Morecambe and back, a pattern which became boringly repetitive after a few weeks.

ENGINEERING AND EXTERNAL CHANGES TO THE P5

In fact, the P5 needed very little change to accommodate the ex-Buick V8 engine. The subframe had to be modified with additional tubular members running fore and aft to carry the new engine mountings. A new exhaust system was drawn up, various ancillaries were repositioned in the engine bay, and the new car was ready to go. Project

This sectioned show engine reveals the V8's rocker gear, pushrods and hydraulic self-adjusting tappets. The engine is actually a P6B type, which differed in the shape of the air cleaner.

131

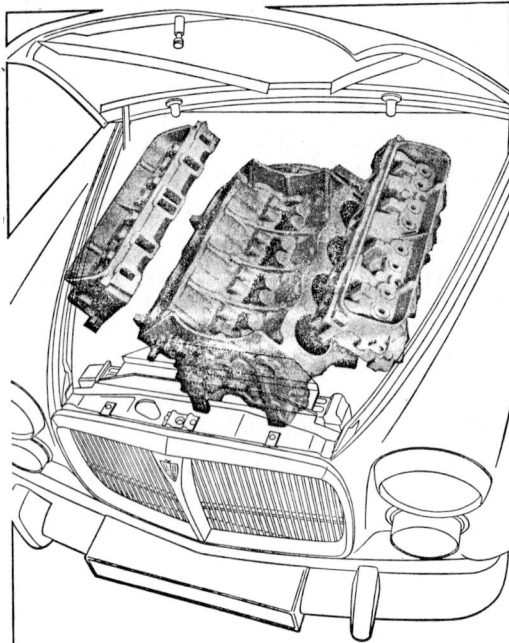

CAST BY **BIRMAL**
FOR ONE OF THE WORLD'S
BEST ENGINEERED CARS.

Stand Nõ 238, Ave. A, 1st Floor.

THE ROVER COMPANY chose Birmal to produce
the sand cast cylinder block and low
pressure die cast cylinder heads in aluminium
alloy for their new V8 3.5 litre engine.

BIRMINGHAM ALUMINIUM CASTING (1903) CO. LTD.
SMETHWICK. WARLEY, WORCESTERSHIRE.

Birmingham Aluminium at Smethwick produced the aluminium alloy castings for the V8 engine, and were proud enough of the fact to place this advertisement in the 1967 London Motor Show catalogue.

Engineer Ken Stansbury remembers that he was given a very restricted timeframe to get the V8-engined P5 ready for production, and that the engine installation was developed on the car and only turned into engineering drawings afterwards, to save time.

The only real problem was that Rover did not have a manual gearbox which was strong enough to handle the V8's torque; neither the elderly four-speed with overdrive

type in the P5 nor the P6's new four-speed was up to it. Work went ahead on the P6 V8 development programme with German ZF manual gearboxes, but Rover decided to put the V8-engined P5 into production without a manual gearbox option. Sales of the 3-litres were in any case heavily biased towards automatics, and the existing Borg Warner type 35 transmission was just able to cope with the V8's torque, although it was quite close to the limit of its capacity. For the V8 cars, Rover asked Borg Warner to modify it slightly to give a control range of P-R-N-D2-D1-L instead of the P-R-N-D-L of the Mk III 3-litres. The new D1 position gave a first-speed start without the need to use kick-down, while D2 gave the traditional second-speed start.

Even though Rover wanted to minimize the engineering changes to the P5s, it was going to be important to make the V8-powered models look different from the 3-litres they would replace. So David Bache's styling department was asked to come up with ideas for a facelift. The budget for this was probably fairly limited, because the V8-powered P5 was expected to have a limited life-span: if all went according to plan, it would be replaced in 1971 by the new P8 saloon. So Bache did what he could.

There could be no question of a fundamental change in the styling, but the money was available to make some small changes to the body pressings. So Bache modified the front wings, borrowing from Jaguar and from BMC's Vanden Plas models the idea of foglamps recessed into the wing fronts. He changed the grille centre section, slimming it down and adding a small, gilded Viking ship badge in place of the multicoloured plastic type. From work his stylists were already doing on the P6 for the US market came slim coachlines which followed the shape of the side trim strip (and of the roof guttering on Coupés), and chromed Rostyle

The 3-litre Coupé on the right appears to have been used as a styling development car for the P5B. The recessed foglamps were retained for production, albeit slightly smaller, but the restyled grille and front bumper were fortunately rejected.

pressed-steel wheels with black-painted panels. To make the cars look lower, Bache blacked out the sill panels and a matching section at the bottom of the front wings. He also added a bright trim strip along each sill, probably to help the cars look longer and therefore lower – although it was a feature he himself had proposed for the original 3-litre back in 1957.

The bumpers picked up squat new overriders with rubber facings, and these exposed the front valance panel and the wing bottoms to such an extent that Bache decided to square them off for neatness. As for the side trim strips, these lost the three pips associated with Mk III 3-litres and were made to embrace neat indicator repeater lamps on the front and rear wings. Ken Stansbury remembers having trouble with these in the prototype stages, as they could not be made to stay on the car! Below the trim strips, and matching up with the

coachline, were new plate-type badges with black backgrounds which bore the model name of '3.5-litre'. Their general style echoed some Triumph badging of the time, and it may be that Leyland – who had owned Triumph since 1961 – asked for a corporate style to be used after Rover merged with them early in 1967. The finishing touch, which was pure stylist's caprice but did hint at the increased performance, was a twin-outlet exhaust tail pipe.

INTERIOR CHANGES

David Bache's department left the existing Mk III-style interior trim largely alone, successfully modernizing it without sacrificing any of its traditional charm. What *The Motor* of 7 October 1967 would later describe as 'the finest London club on wheels' was brought up to date by means of a neat centre console between the driver's and front passenger's seats. The catalyst for this styling change had actually been a decision to move the automatic transmission selector lever from the steering column to the transmission tunnel, where the P5B installation paralleled the one pioneered on the 2000 Automatic in 1966. However, the new console was a stroke of genius which allowed the pull-out picnic tray and a large central ash-tray (once again a P6 item) to be grouped together in a single integrated piece of design.

In only one respect was the P5B's passenger cabin inferior to the one of the car it replaced, and this was clearly a result of cost-cutting in the manufacturing area. Where the carpet of the Mk III 3-litre had always matched the colour of the rest of the trim, the carpets of P5Bs came in just two colours: Mortlake Brown and Silver Grey. The brown toned in well with the light and dark brown trim colours, but the grey never looked quite right with the Mulberry (plum)

and Ebony (black) colours it was supposed to complement.

THE PILOT-PRODUCTION CARS

The first pilot-production V8-engined P5 Saloon was built in February 1967, and the first Coupé followed at the beginning of March. These and the few further examples built before the factory's summer shutdown in August – when the 3-litre assembly line was converted to build the new models – had pre-production engines numbered in the Engineering Department's '2158/xx' series. When the factory opened again in September, the 3-litres were no more and the cars which came down the single assembly line in the South Block were the new P5B models, the B of their designation standing for Buick, whose engine had given the ageing design a new lease of life.

This side view of the styling development car suggests that David Bache had been casting an eye in the direction of the Mercedes 600 limousine; the general shape of the bumpers, the bright wheelarch mouldings and the heavy chrome strip along the door bottoms all hint at the big German limousine. However, they made the P5 look fat and overweight, and were not adopted for production.

8 The 3.5-Litre and 3½-Litre Saloons and Coupés, 1967–1973

The most important difference between the P5B and its 3-litre predecessors was of course the improved performance offered by the V8 engine. Although the V8 was never as quiet at idle as the old straight-six had been, its exhaust note was a discreet burble rather than the rorty rumble so often associated with V8 engines, and it certainly delivered the goods. Where a Mk III 3-litre automatic had taken a leisurely 17 seconds to reach 60mph from rest and had peaked at about 104mph, a 3.5-litre would storm up to 60mph in 12.5 seconds and would go on to a maximum of 110mph. This was performance very similar to that of the much-praised 2000TC – then the fastest Rover on sale – and although it was not in the Jaguar class, it did make the P5B a much more attractive proposition than the Mk III 3-litre had been. There was more, too: the lighter engine reduced the car's tendency to understeer and thus made the increased performance that much more useable on

Very few 3.5-litre Rovers were put on the road with the E-suffix registrations current for the first half of 1967. RXC 436E was one of the very first Saloons built and appeared in all the early press material. In later years, the original pictures were re-used for publicity purposes, with the number plate air-brushed to read RXC 436K!

Rover 3.5-litre Saloon and Coupé (1967–1968)
Rover 3½-litre Saloon and Coupé (1968–1973)

Layout
Monocoque bodyshell with front subframe bolted in place. Five-seater saloon, with front engine and rear wheel drive.

Engine

Type	215/8
Block material	Aluminium alloy
Head material	Aluminium alloy
Cylinders	Eight, in 90-degree vee
Cooling	Water
Bore and stroke	88.9mm × 71.12mm
Capacity	3,528cc
Main bearings	Five
Valves	Overhead, two per cylinder
Compression ratio	10.5:1
Carburettor	Two SU type HS6 (1.75in)
Max. power	184bhp gross (160.5bhp net) at 5,200rpm
Max. torque	226lb.ft at 3,000rpm

Transmission
All cars had automatic transmission with a torque converter.

Internal gearbox ratios
Three-speed automatic (Borg Warner type 35)

	(Final drive 3.54:1)
Top	1.00:1
Intermediate	1.45:1
First	2.39:1
Reverse	2.09:1

Suspension and steering

Front	Independent, with wishbones, laminated torsion bar springs and anti-roll bar
Rear	Semi-floating axle with progressive-rate semi-elliptic leaf springs
Steering	Burman recirculating ball type with variable ratio and power assistance
Tyres	6.70 × 15 crossply
Wheels	Five-stud pressed-steel Rostyle, chromed and painted
Rim width	5in

Brakes

Type	Discs at the front and drums at the rear, with servo assistance
Size	Disc diameter 10.75in
	Drum diameter 11in

Dimensions (in/mm)	
Track, front	55.3 (1,405)
Track, rear	56 (1,422)
Wheelbase	110.5 (2,807)
Overall length	186.5 (4,737)
Overall width	70 (1,778)
Overall height	60 (1,524) (Saloon)
	58 (1,473) (Coupé)
Unladen weight	3,498lb (1,587kg) (Saloon)
	3,479lb (1,578kg) (Coupé)

The major new interior feature on the P5Bs was a centre console, incorporating the transmission control lever, an ashtray, the cigarette lighter and switches for the foglamps and heated rear window. The tool tray which doubled as a picnic table formed a neat centrepiece at the top. This early publicity picture shows an A-suffix car with the original knob on the transmission selector; on later cars, the knob had a shaped cut-out on the top to prevent accidental operation of the detent button.

twisting roads, while it also reduced fuel consumption from the typical 16–17mpg of an automatic 3-litre to around 20mpg.

Nevertheless, these changes did not turn the Saloons and Coupés into best-sellers overnight. What they did do was to rejuvenate the cars and keep them alive for a little longer, until they could be properly replaced. There was an initial spurt of customer interest – Rover had planned for 85 cars a week and had to double that to meet demand – but sales began to slide at the turn of the decade. Those of the Coupés tailed off faster than those of the Saloons, which saw a last-minute

upturn that was at least partly attributable to Government and military batch orders. By the end, however, the P5B had become a very low-volume model indeed, with weekly production averaging out at about 50 cars.

Export sales of these cars were also slow. The P5Bs never did make it to the USA, where the Rover Motor Company of North America was by this stage concentrating on the P6 models, but they continued to sell slowly in some countries of the old Commonwealth where British tradition and values were appreciated. In northern Europe only The Netherlands showed much interest, and

The Coupé once again proved most popular in two-tone colour schemes, and Rover stands at motor shows regularly featured cars with Admiralty Blue lower panels and a Silver Birch roof. This car was almost certainly finished in those colours, with the light Buckskin interior which went so well with them. Despite the impression given by this picture, the Coupé was generally an owner-driver car; those who had chauffeurs normally preferred the extra room in the back of the Saloon.

Sumptuous comfort in the rear of a 3.5-litre Coupé, although the front seats have been pushed forward on their runners to make the legroom look more acceptable! More legroom did become available when thinner seat backs were introduced on the B-suffix cars in 1968.

3.5-litre becomes 3½-litre

The P5Bs were known as 3.5-litre models only for their first season of production. In April 1968, after the cars had been on sale for just over six months, Rover introduced the V8-engined edition of the P6 saloon, which sported '3500' badges to make clear that it was related to the 2000.

Even though Rover tried to make the distinction between the two V8 cars clear by insisting that the 3500 was called a Three Thousand Five, it soon became obvious that there was some confusion over model-names. So for the 1969 season, beginning in September 1968, the 3.5-litre was renamed a 3½-litre. It nevertheless retained its original '3.5-litre' badges.

Rover might just as well not have bothered. Confusion about which car was which still reigned by the mid-seventies, and customers at the spares counter of a Rover dealership were always likely to be asked whether the parts they wanted were for a '2000 shape' or 'the big old one'!

a few cars were sold in Scandinavia. By this stage, however, the P5B was seen very much as a survivor from a bygone age, and as such it appealed to a very limited clientele.

ON SHOW

Rover announced the new cars to the press in September 1967, and several reports had been published by the time the Earls Court Motor Show opened on 18 October. However, that show was the first opportunity the general public had to see the 3.5-litre cars in the metal. The Rover stand was number 123, and the company exhibited three P5Bs alongside four examples of the P6 range. On a plinth was a Coupé in perhaps the most attractive of all colour schemes – Silver Birch over Admiralty Blue with a Buckskin interior – while a second Coupé was finished in Burnt Grey over White with Mulberry trim

The P5Bs had a gold-coloured grille badge which looked good when new but was extremely difficult to keep clean.

and the single Saloon was in Arden Green with Saddle Tan trim.

Paint and trim colours, Rover 3.5-litre and 3½-litre Saloon and Coupé

1968–1973 model years
There were seven standard paint colours and four standard interior colours which were matched with just two colours of carpet. Saloons were always single-tone unless to special order, and Coupé models were available with seven two-tone paintwork combinations. The standard combinations were:

Single tone	Upholstery	Roof on two-tone Coupés
Admiralty Blue	Buckskin, Mulberry or Saddle Tan	Silver Birch
Arden Green	Buckskin or Saddle Tan	Silver Birch
Bordeaux Red	Buckskin, Ebony, Mulberry or Saddle Tan	Silver Birch
Burnt Grey	Buckskin, Ebony, Mulberry or Saddle Tan	Silver Birch
Rover White	Ebony, Mulberry or Saddle Tan	Burnt Grey
Silver Birch	Ebony, Mulberry or Saddle Tan	Burnt Grey
Zircon Blue	Buckskin or Saddle Tan	Silver Birch

Carpets were in Mortlake Brown with Buckskin and Saddle Tan upholstery, or in Silver Grey with Ebony and Mulberry upholstery. **Coachlines** on single-tone cars were white with all colours except Rover White and Silver Birch, which had black coachlines. On two-tone Coupés, the coachline on the roof was in the same colour as the lower body, while the bodyside coachline was in the same colour as the roof. (*Note*: The details of coachline colours provided in a *Rover Service Newsletter* dated September 1967 are incorrect but may have represented a planned specification.)

Examples of the P5B would feature on the Rover stand at Earls Court every year up to and including 1972, but the cars would never again take centre stage. At the 1968 show, all the excitement centred on the recently-announced P6B with its V8 engine, and just two years later the Range Rover would steal the limelight.

ON TEST

The first press demonstrator of the 3.5-litre Saloon was registered as RXC 436E, and *Autocar* magazine borrowed it for the first published road test of the new models, which appeared in the issue dated 28 September 1967. The 3-litre, the testers concluded,

> was generally considered – particularly in its automatic form – to be rather staid and stodgy. Now with the vee-8 under the bonnet it has made such gains in performance and economy that it has gone ahead of most of its competitors, and has taken on a different image by being so much faster and more manageable. It is also exceptionally good value for money.

The road-test staff took the car on a 3,000-mile trip through Scandinavia and continental Europe, as well as testing it to a 108mph maximum – somewhat short of the 115mph claimed by Rover. They found the performance improvement most noticeable at speeds above 80mph, where the 3-litre tended to tail off, and appreciated the improved handling and braking which the lighter engine brought. 'In one respect,' however, 'the new car is not as good as before. There is now some wander when running straight ... but the stability of the Rover is still very good indeed.' There was also another problem, which was not reported in the road test but came to light in a

Distinguishing features at the rear of the P5Bs were the more widely spaced letters of the Rover name, the 3.5-litre plate badge (accompanied by the familiar 'Coupé' script where appropriate), and twin exhaust tailpipes.

report on a used 3½-litre Coupé in 1972: the transmission had 'an alarming habit, which we met on the Road Test 3½, of making a very noticeable 'bonk' while changing down as one coasts to a halt at traffic lights.'

John Bolster of *Autosport* magazine had been a staunch supporter of the 3-litre models over the years, and he enthused about the road-test Saloon registered TXC 687F in the magazine's issue dated 29 September 1967. 'The new 3.5-litre Rover,' he summarized, 'is a phenomenally quiet car of great refinement. It is a worthy successor to a long line of Rover 'sixes' and adds a useful turn of speed to their many virtues.'

Like *Autocar*, Bolster could not better a timed 108mph, but serene cruising was possible at 90mph, and the lighter front end enabled the car to corner better than its six-cylinder predecessors. The new V8 engine 'makes light of the considerable weight which is a penalty of such luxury,' and using the D1 gear range made the car feel 'really lively, accelerating briskly past slower cars

and changing into top just below 70mph.'
The P5B was a car of

> great dignity, in spite of its raffish new
> wheels. Nevertheless, quite a few casual
> passers-by were critical of these wheels, and
> I think the makers would be wise to offer an
> alternative type to their more conservative
> customers.'

The *Motor* testers were very impressed with
the Coupé that they borrowed and reported
on in the magazine's issue dated
7 October 1967.

> So many who regard this Rover – with its
> wood-panelled interior and four thick,
> leather armchairs – as being the finest
> London club on wheels, will now have to
> accept it as being the fastest as well.

The Coupé achieved 110mph on the MIRA
banking, and even in the more leisurely D2
range of its automatic transmission was
much faster than the 3-litre automatic it
replaced.

> The figures show how the big Rover has

been lifted into the high-performance class
by its new light-alloy 160.5bhp V8 engine.

Quietness remained a virtue:

> the Rover is so well insulated from all
> sources of noise that a Jeeves-like calm
> nearly always prevails; even when the
> engine is at is most discreetly agitated it
> emits nothing more than a purposeful hum.

It was:

> quite possible to cruise at well over 100mph
> with the radio comfortably audible and the
> engine sounding quite unstrained.

The 3.5-litre car also handled much better
than its 3-litre predecessor, and

> roll, understeer and tyre squeal have all
> been reduced.

And, of course, there were other delights to
be sampled:

> what distinguishes the Rover from its com-
> petitors is the number of undeniably

*This centre console shows the
standard switch configuration,
with the cigarette lighter
flanked by heated rear window
switch on the left and foglamp
switch on the right. The
gearshift pattern is the later
D-2-1 type and the selector
lever has the later knob with
a shaped cutout. Not standard
is the switch alongside the
selector, which operates a
Hollandia electric sunroof.*

141

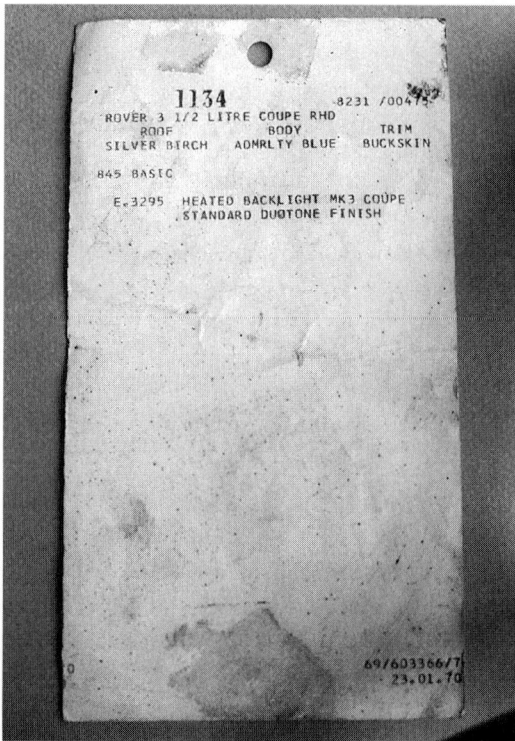

Every P5 and P5B had one of these ... once. It is the line ticket, giving the line build number (1134 in this case) and the specification of the car. Note that the actual car number was not stated on the build ticket. The E-numbers refer to optional extras, in this case only a heated rear window. These tickets are often found tucked behind the dashboard or under the headling.

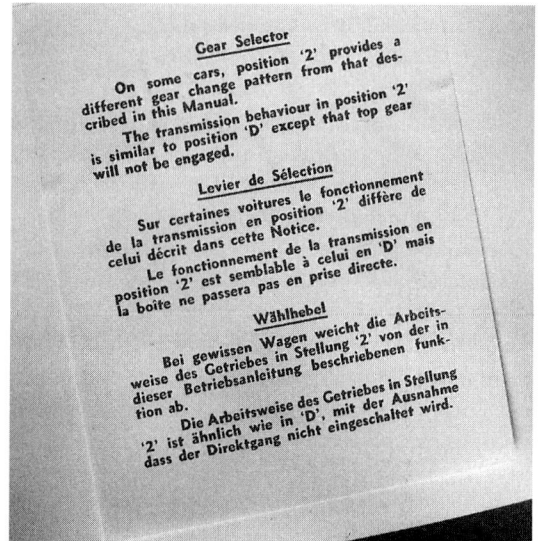

This is a curiosity from early 1970. When a batch of 3½-litres had to be fitted with a non-standard automatic transmission, the owner's handbooks which went with them were modified with this sticker.

practical items which afford almost child-ish pleasure every time they are used:

items like the front and rear picnic trays, the adjustable armrests and the front and rear ashtrays with their illuminated lighters.

Nevertheless, the *Motor* team did concede that the body styling was by now 'rather old-fashioned'. The layout of the minor controls also came in for criticism, as it was impossible for the driver to reach some of them while wearing his safety belt. There was some suspension harshness on bumpy roads, and some of the testers criticized the lack of 'feel' in the power steering, which could give a feeling of insecurity at speed on winding country roads or when cruising on the straight at over 100mph. After nine years in production, perhaps the P5 was at last beginning to show its age.

Like John Bolster in *Autosport*, *Motor Sport's* Bill Boddy wondered in November 1967 whether the new wheels would meet general approval. His test car was the Saloon registered RXC 436E, and he was very much in favour of its new-found performance:

the new engine has considerably increased the performance of the car, so that a top speed of around 110mph and a standing-start ¼-mile accomplished in under 18½ seconds has completely dispelled the staid aspect of the biggest car from the Rover factory.

142

He also pointed out that the car was quieter than the Jaguar 420, with which it was bound to be compared.

However, the design was definitely beginning to show its age. The suspension was too lively over bad roads, the car still rolled a lot, and its steering was not as accurate as Boddy might have liked. On the credit side, though, it did handle better than a 3-litre and could be driven quickly without difficulty. 'I would describe it as similar to a Rolls-Royce Silver Cloud to drive,' Boddy summarized; 'outdated but highly satisfying.'

Thereafter, magazine reports on the P5B were few and far between. Rover did keep examples on the press demonstrator fleet, but the car changed so little during its five years of production that there was nothing new to write about. However, *Autocar* tested a used 1970 3½-litre Coupé in its issue of 23 March 1972, and it is interesting to see there the way in which the car was viewed

3.5-litre and 3½-litre Saloon: commission numbers and production figures

1968 3.5-litre Saloon	840-00001 to -02433	Home market	(2,433)
	841-00001 to -00104	Export, RHD	(104)
	843-00001 to -00116	Export, LHD	(116)
			Total: 2,653
1969 3½-litre Saloon	840-02434 to -04000	Home market	(1,567)
	841-00105 to -00318	Export, RHD	(214)
	843-00117 to -00234	Export, LHD	(118)
			Total: 1,899
1970 3½-litre Saloon	840-04001 to -05756	Home market	(1,756)
	841-00319 to -00509	Export, RHD	(191)
	843-00235 to -00354	Export, LHD	(120)
			Total: 2,067
1971 3½-litre Saloon	840-05757 to -07493	Home market	(1,737)
	841-00510 to -00569	Export, RHD	(60)
	843-00355 to -00390	Export, LHD	(36)
			Total: 1,833
1972 3½-litre Saloon	840-07494 to -08783	Home market	(1,290)
	841-00570 to -00639	Export, RHD	(70)
	843-00391 to -00416	Export, LHD	(26)
			Total: 1,386
1973 3½-litre Saloon	840-08784 to -10341	Home market	(1,558)
	841-00640 to -00723	Export, RHD	(84)
	843-00417 to -00437	Export, LHD	(21)
			Total: 1,663
			Total: 11,501

just a year before it went out of production. This was still 'the sort of car which Britain can produce better than any other country,' and, particularly in Coupé form, it was 'a completely individual design; you either like it or have no interest in it.'

Rover and British Leyland

When the P5 entered production in 1958, the Rover Company Ltd was an independent motor vehicle manufacturer, just as it had been since it changed its name from the Rover Cycle Company in 1906. Its principal factory and administrative headquarters were at Lode Lane in Solihull (which is now occupied entirely by Land Rover Ltd), and had been since 1945 when the company had moved there from its bomb-damaged original home in Coventry.

The Rover Company had become both well regarded and highly prosperous by the middle of the sixties, thanks in a very large measure to the worldwide success of the Land Rover after its introduction in 1948. Expansion was on the agenda: Managing Director William Martin-Hurst was particularly keen to take Rover successfully into the US market, and during 1965 the company bought out the old-established Alvis concern. Plans were soon laid to develop new Alvis models, using existing Rover mechanical components.

However, it was also in 1965 that the British Motor Corporation bought out Pressed Steel, Rover's long-standing and only body suppliers. Pressed Steel guaranteed to fulfill existing contracts, but there was no telling what might happen in the future. Jaguar quickly sought to guarantee its supplies of Pressed Steel bodies by merging with BMC in 1966 to form British Motor Holdings, and this forced Rover to seek shelter with the Leyland Motor Corporation in the early months of 1967. Leyland were best known as makers of trucks and buses, but had already bought out Standard-Triumph International and thus owned body plants at Bordesley Green (the old Mulliners of Birmingham premises) and Tile Hill (originally built by Fisher and Ludlow). The P5B was thus the first new Rover to be announced after the company lost its independence.

However, the situation would change again soon after that. Harold Wilson's Labour Government had begun to fear that the increasing success of big foreign motor manufacturers like Volkswagen, General Motors and Ford would come to squeeze the smaller makers out of the market, and was understandably anxious to protect jobs in the British motor industry against this threat. So, believing that a large British-owned motor manufacturing combine would be the best defence, the Government arranged a shotgun marriage between British Motor Holdings and the Leyland Motors Group early in 1968. Rover thus became part of the new British Leyland Motor Corporation.

For the first few years, Rover was left to carry on very much as it had before. However, in the early seventies, BL started to rationalize the vast assortment of manufacturers and suppliers which had come under its umbrella, and early in 1971 it merged the engineering departments of Rover and Triumph. This was the beginning of the end for Rover as it had been. The Rover Company now became Rover-British Leyland UK, Ltd, as a subsidiary of Rover-Triumph, and overall control moved away from Solihull. The P8 luxury saloon which Rover had planned should replace the P5B was now seen as a dangerous competitor for Jaguar's XJ6, and so BL cancelled it in March 1971. Several million pounds had already been spent on its development and on production tooling for its planned launch in autumn 1971.

So it was that the P5B was kept in production for a couple of years longer than Rover had originally planned. But in its last two years, it was being built by a disillusioned and demoralized workforce who no longer took the same pride in their products as they had under the old Rover Company. As a result, the quality of the final P5Bs was not as high as that of earlier examples.

RIVALS

'The only competitive car which really worried us was the Jaguar 420,' admits Project Engineer Ken Stansbury. At the 1967 Motor Show which saw the introduction of the V8-engined Rovers at £1,625 before purchase tax for the Saloon and £1,705 for the Coupé, the 420 cost a very competitive £1,615 with overdrive or £1,678 with automatic transmission. However, Jaguar took it out of production during 1968, and replaced it with the acclaimed XJ6. In 4.2-litre non-overdrive form this was carefully priced at £1,762 before purchase tax, mid-way between the 3½-litre Saloon at £1,700 and the Coupé at £1,775. In practice, however, most XJ6 buyers spent the extra on overdrive or on an automatic transmission, which put the Jaguar's price well above that of the Rover Coupé and thus largely removed it as a competitor. The 2.8-litre XJ6 was quite a lot cheaper than the Rovers, at £1,563 with automatic transmission in October 1968, but it was never a well-liked car and probably did not present a serious threat.

Various other cars came up against the Rover on price between 1967 and 1973, but none of them really presented a viable alternative on the home market. The Vanden Plas 4-litre R, which disappeared in 1968, was actually rather more expensive than the Rover. The 2-litre BMWs and the remarkable NSU Ro80 were competitive on price but not on luxury features, and the Volvo 164 which just undercut the P5Bs on price was also rather Spartan by comparison. Overseas, however, where the Rover's price was often inflated by import taxes, the car simply could not compete with well-specified Mercedes or Citroëns at similar prices. It was for that very reason that export sales were fairly poor.

THE SUNROOF OPTIONS

The Webasto fabric sunroof, approved by Rover as a dealer-fitted option for the 2000 saloons as early as 1963, did not become optional for the P5s until June 1968. Whether this was because Webasto had not developed a suitable installation before then is unclear.

A small number of cars had sliding metal sunroofs, but these never became a regular option. This one is on a 1971 Coupé.

Optional extras available for the 3.5-litre and 3½-litre models

Badge bar
Bench rear seat for Saloons (no-cost option)
Dipping interior mirror
Door mirror (from January 1972)
Extension speaker for radio (fitted under rear parcels shelf), with balance control
Floor mats, Beige nylon fur
Floor mats, Charcoal Grey rubber
Floor mats, rubber link type
Foot pump
Head restraints, front (Saloon and Coupé) and rear (Saloon only)
Headlamp lens masks in amber, for continental touring
Heated rear window
Laminated windscreen
Mud flaps, front and rear
Radio (Radiomobile Medium Wave/Long Wave; Radiomobile Short Wave/Medium Wave alternative from June 1970 to October 1972; Radiomobile Medium Wave alternative from May 1971 to October 1972)
Radio aerial (always roof-mounted before July 1971, when a wing-mounted alternative was introduced)

Roof rack (Saloon only to August 1970; both models thereafter)
Seat belts (Irvin static type) for front and rear; inertia-reel front belts available as alternatives from May 1971 and fitted as standard from January 1972
Snow chains
Special paint finish or interior trim to customer's requirements
Sundym tinted glass (from August 1969); with laminated windscreen from October 1971; with heated rear window as standard from April 1973
Towbar
Two-tone paint (Coupé only)
Webasto fabric sunroof (from June 1968)

Note: Wing mirrors were never listed as an option for the P5B models and were never fitted to the factory's demonstrator vehicles. Many cars were nevertheless fitted with them when new, presumably at customer request. Rover also approved an air-conditioning installation by Hooper Motor Services in London from June 1969, but this was never advertised in showroom literature and remained very rare indeed.

The P6s could also be fitted with a Hollandia sliding steel sunroof after October 1968 (although very few were), and in the spring or summer of 1972 a number of P5Bs, both Saloons and Coupés, were fitted experimentally with Hollandia sunroofs. Some had manual operation and some electric; more precise details are lacking, but it seems probable that around ten or a dozen cars were so fitted. The known ones were registered by the Rover Company with K-suffix numbers in the Solihull XC series. Examples are CXC 77K (840-09070D), a Bordeaux Red Saloon with the manual roof registered on 13 June 1972, and XXC 509K (845-06271D), a Silver Birch over Admiralty Blue Coupé which was probably the 1971

Earls Court Show car and was fitted with an electric roof after sale by Rover in July 1972.

PRODUCTION CHANGES

Serious development work on the P5B ceased once the car had entered production, but Project Engineer Ken Stansbury was kept busy with a stream of running changes. There were three suffix-letter changes marking major design modifications, thus creating four main variants of the cars. Suffix-letters, incidentally, were in capital letters on the P5Bs, whereas they had been in lower-case letters on the 3-litres. The first cars had A-suffix numbers, and lasted until August 1968.

The B-suffix cars lasted just a few months until the C-suffix models were introduced in two stages; all types except home market Saloons changed in December 1968, but it was February 1969 before the 840-series cars picked up the C-suffix modifications. The final suffix change was to D in September 1969, and all the subsequent Saloons and Coupés had this suffix.

The change from an A-suffix to a B-suffix was brought about by the introduction of a revised automatic transmission. The original 1968-season models had D2, D1 and L settings in their transmissions, D1 giving all three gears and D2 cutting out bottom gear to give the second-speed start which offered smoother operation in stop-start traffic, while L locked the transmission and prevented it from changing up in the lower two gears. For 1969, however, the selector gate was changed to give D, 2 and 1; in this case, D gave all three gears, 2 gave second only, and 1 gave first only. In practice, driver override of the transmission was greater because

3.5-litre and 3½-litre Coupé: commission numbers and production figures

1968 3.5-litre Coupé	845-00001 to -01610	Home market	(1,610)
	846-00001 to -00080	Export, RHD	(80)
	848-00001 to -00089	Export, LHD	(89)
			Total: 1,779
1969 3½-litre Coupé	845-01611 to -03624	Home market	(2,014)
	846-00081 to -00240	Export, RHD	(160)
	848-00090 to -00230	Export, LHD	(141)
			Total: 2,315
1970 3½-litre Coupé	845-03625 to -04944	Home market	(1,320)
	846-00241 to -00368	Export, RHD	(128)
	848-00231 to -00333	Export, LHD	(103)
			Total: 1,551
1971 3½-litre Coupé	845-04945 to -06405	Home market	(1,461)
	846-00369 to -00430	Export, RHD	(62)
	848-00334 to -00359	Export, LHD	(26)
			Total: 1,549
1972 3½-litre Coupé	845-06406 to -07410	Home market	(1,005)
	846-00431 to -00465	Export, RHD	(35)
	848-00360 to -00391	Export, LHD	(32)
			Total: 1,072
1973 3½-litre Coupé	845-07411 to -08195	Home market	(785)
	846-00466 to -00502	Export, RHD	(37)
	848-00392 to -00402	Export, LHD	(11)
			Total: 833
			Total: 9,018

second gear could be locked, whereas 'L' on the original transmission did allow a change down from second into first.

For 1969, the front seats also had thinner backs and softer cushions, the first to give more legroom in the rear and the second to give greater comfort. With these thinner seats came revised headrests, now with a straight metal support pillar instead of a cranked one. In addition, the B-suffix cars had a larger rear view mirror instead of the much-criticized diminishing mirror which the A-suffix models had inherited from the 3-litres.

The modification which brought about the change to suffix C was the introduction of an automatic choke. The AED, or Automatic Enrichment Device, was an SU component which had been developed initially for the P6B 3500S models sold in the USA after June 1969. In theory, it should have enhanced the luxury specification of the P5B, but in practice it gave the Rover engineers nightmares for the next few years, and in the end the company was obliged to introduce a conversion kit to manual choke. (The official line, when this was introduced in March 1974, was that it was a temporary expedient to overcome supply difficulties of replacement AEDs, but there is absolutely no doubt that it was actually an admission that the AED could not be made to work satisfactorily.) The AED caused a variety of problems, including preventing a cold engine from starting, preventing a hot engine from starting, and remaining in operation after the engine had warmed up and thus leading to high fuel consumption and stalling. Many owners simply removed the offending device or by-passed it as soon as they could.

It was a modified downshift cable for the automatic transmission which brought about the last suffix change, to D. Subsequent changes were fairly minor, although it

is worth mentioning that a thief-proof coil was fitted after December 1970. This required an armoured cable running from coil to ignition switch, but the cable could not be persuaded to go through the twists and turns needed to reach the existing switch. So Rover relocated the switch on an extension to the left of the steering column. Sadly, the thief-proof ignition proved neither thief-proof nor reliable in the long term, although it remained in the specification of the P5B until production ended. In later years, however, most owners by-passed the system and fitted standard coils.

Two other areas gained more than their fair share of attention during the P5B's production life, and these were the brakes and the power steering box. Owners were clearly making use of the extra performance afforded by the V8 engine, because Rover became worried about premature brake pad wear. So first a harder pad option was introduced, and then harder pads were standardized. The power-steering box, meanwhile, suffered from leaks. A modification failed to cure the difficulty, which remained with the car until the end of production. Subsequently, it became clear that a fault in the tooling for one of the steering box's rubber seals had resulted in a minor imperfection which was responsible for the leaks.

There was also one production change – or perhaps production anomaly – which has so far defeated every attempt to pin it down. A number of the later Coupé models, probably all D-suffix cars, had black Viking ship badges on their rear-window pillars instead of the more familiar silver ones. However, these badges appear to have been fitted more or less at random, and remained the exception rather than the rule. Without some very detailed research or a stroke of luck, the precise details of this chapter of the P5 story seem destined to remain elusive!

Rover 3.5-litre and 3½-litre Saloon and Coupé: production changes

Note: All dates given for changes where no chassis number is quoted must be treated as approximate. The dates are taken from issues of *Rover Service Newsletter*, which usually reported changes between one and four months after they had actually been made on the assembly lines. Dates given for changes where a chassis number is quoted are exact for home market models and reflect the date into Despatch (i.e. the date the car was transferred from the assembly lines to the Despatch Department) given in records held by the British Motor Industry Heritage Trust.

Date	Home, Manual	Home, Auto	Other	Remarks
September 1967	3.5-litre Saloon and Coupé introduced.			
October 1967				Revised HT lead routing to prevent cross-firing.
December 1967	840-00735A	845-00539A		Heated rear window with enlarged heated area and yellow elements replaced laminated type with 'invisible' elements. (Old type still fitted to some cars after this date.)
	840-00746A	845-00848A		Rubber grommet in air cleaner body now cemented in place.
January 1968	840-00880A		843-00009A	Two-piece rear seat valance trim introduced; retro-fit possible.
	840-00892A	845-00611A	843-00009A	African Walnut wood mouldings replaced Cherrywood.
				Brass-bodied water temperature transmitter unit replaced steel and aluminium type to prevent corrosion; retro-fit possible.
March 1968				Recommendation to retard ignition timing to TDC to allow 96 octane fuel to be used when touring abroad (all cars for Belgium, France and Holland had TDC timing as standard from August 1968).
				Second-skim machining operation on sump joint face of cylinder block discontinued, from engine number 840-02696A.
April 1968				Body-colour identification labels added.
				Modified oil screen housing assembly introduced to increase clearance between housing and sump, from engine number 840-01349A; retro-fit possible.
June 1968				Up-rated starter motor introduced (identifiable by

two terminals instead of four on solenoid and by number 26266 in place of 26254B stamped on casing); retro-fit possible.

Driver's door mirror option introduced; retro-fit possible.

August 1968

Modified forward sun gear assembly and front clutch cylinder in automatic transmission, from transmission number 3EU 5969.

1969 model-year

September 1968   Suffix B cars introduced

Models now known as Rover 3½-litre Saloon and Coupé

Petrol filter repositioned with centre flange above clip for better support; petrol pipes modified to suit.

D-2-1 transmission (serial prefix 5FU on red plate) replaced D2-D1-L type (serial prefix FU on blue plate).

Trico windscreen washer reservoir replaced Lucas type.

Inhibited ignition and starter switch.

New oil pressure gauge on Coupés, with improved accuracy in low pressure range; retro-fit possible.

Front seats with softer cushions and thinner squabs.

Straight pillar replaced cranked type on optional front head rests; stop added to headrest pillar retainer.

Larger rear view mirror.

Smaller underbonnet lamps on Coupés (date approximate).

Lucas 8FL 'no delay' flasher unit fitted.

October 1968

Recommended replacement interval for cartridge fuel filter reduced from 20,000 miles to 10,000 miles.

	Fibreglass insulation pad deleted from inlet manifold.
November 1968	Harder brake pads with Ferodo 2430F lining material introduced as optional service replacements.
December 1968	Engine oil capacity reduced from 9 pints (5 litres) to 8 pints (4.5 litres) to prevent overfilling; dipstick 'High' mark changed to suit.
	Electrical kit now available for trailer flashers (early examples contained the wrong flasher unit as supplies were short).
	Suffix C Coupés and Export Saloons introduced.
	AED introduced on Export Saloons and on all Coupés.
January 1969	Larger water passages in timing cover and water outlet elbow, to increase by-pass water flow.
	Improved end cover, seal and seal retainer for PAS box, to prevent fluid leakage.
February 1969	Suffix C home market Saloons introduced.
	AED introduced on home market Saloons.
April 1969	25-amp fuse now specified for heated rear window circuit; recommended as retro-fit.
June 1969	Approval given to air conditioning system fitted by Hooper Motor Services of London.
August 1969	Extended dipstick assembly (with white polypropylene hand grip) to improve accessibility, from engine number 840-07557B.

1970 model-year

September 1969	Suffix D models introduced, with engine suffix C
	Shorter downshift cable for automatic transmission, to maintain consistent line pressure setting (downshift cable supported vertically at a cylinder head bracket and engine breather outlet hose secured at a clip mounted to carburettor fixing).

	840-04503D	845-03375D	841-00296D 843-00227D 846-00228D 848-00210D	Inertia-reel front safety belts introduced.

October 1969 — Approval granted to Dunlop CW 44 6.70 × 15 winter tyres; maximum speed of 95mph to be observed.

Modified gearchange indicator plate retaining spring with split pin fixing.

New BC post trim assembly to accommodate inertia-reel front safety belts.

Safety belts with press-button release replaced belts with over-centre buckle.

Modified dipstick with repositioned 'High' and 'Low' level markings (to allow for adverse tolerance variations in oil filter tube); new dipstick identified by black lettering on cap.

December 1969 — Improved condenser with closer manufacturing tolerances, to provide longer life for distributor contact points.

Brake pads with Ferodo 2431F lining material introduced in place of 2424F type, to improve pad life.

January 1970 — Modified front gearbox band, now with ribbing; retro-fit possible.

February 1970 — Temporary fitment of automatic transmission with different gear change pattern: '2' allowed changes between first and second speeds instead of holding second.

Relocation of transmission breather tube recommended to cure oil ejection problems; modified tube not incorporated into production cars.

Introduction of service kit to allow transmission with short downshift cable to be fitted to earlier cars with long downshift cable.

June 1970	Engine rocker shafts with improved nickel-plated finish, to reduce wear.
	Self-adjusting front brake band servo fitted to automatic transmissions.
	Distance piece introduced to enable late front headrests to be fitted to suffix A cars.
	Ignition coil now fitted with one male and one female connector, to prevent reversal of polarity; adaptor connector available to allow retro-fit.
August 1970	New ignition harness with longer caps for distributor connections; these caps of matt black Hypalon instead of shiny black PVC.

1971 model-year

September 1970	Balanced engine fan assembly with straight blades replaced curved-blade fan to minimize the possibility of blade fracture; retro-fit possible.
	New clip to retain kickdown cable to bellhousing.
October 1970	Brake discs now retained to hub by bolt and plain washer instead of set screw and spring washer, to prevent fixings working loose; retro-fit possible.
	Adhesive balance weights introduced for road wheels.
November 1970	Tool kit jack changed to Three Thousand Five type with rubber bung at top; clips on tool board in boot modified to suit.
December 1970	Thief-proof ignition coil fitted to all home market models, to meet new UK regulations effective 1 January 1971; ignition switch moved to steering column shroud to suit.
	Key numbers deleted from car serial number plate.
February 1971	Aluminium die-cast hub for fan and water pump pulley replaced cast iron type, from engine

				number 840-10542C; retro-fit possible.
March 1971				Improved seal between AED hot air pick-up and left-hand exhaust manifold, from engine number 840-13188C.
				Gearchange indicator plate with brush masking replaced plate with metal masking; gear selector lever also modified with smaller ball to facilitate gear selection.
May 1971				Cruciform oil seal packing on rear main bearing cap replaced straight type, from engine number 840-14168C.
				Pistons with W-slot design introduced, to increase control over piston expansion; modified crankshaft to compensate for increased weight of pistons.
				More flexible top compression ring for pistons, to prevent fracture.
				Improved AED introduced, type number AUH 300.
				267-series automatic gearboxes (with blue plate) replaced 9FU-series (with Grey plate).
				Starter inhibitor deleted from ignition switch.
June 1971				Revised torque converter and housing to minimize the possibility of heterodyne at high speeds and water ingress to starter motor.
				Passport to Service book replaced *Owner's Maintenance Schedule Book* in literature pack, with effect from 1 June 1971.

1972 model-year

October 1971	840-08317D	845-06406D	841-00570D 843-00391D 846-00431D 848-00360D	Modified automatic transmission, with serial number prefix 303 on yellow identification plate; retro-fit possible as a complete unit.
January 1972	840-08656D	845-06686D	841-00581D 843-00402D 846-00438D 848-00369D	Don 227 brake pad material replaced Ferodo 2431F, to eliminate possibility of brakes pulling.

February 1972				Modified engine oil dipstick (with white plastic grip) and tube to improve accuracy of readings, from engine number 840-16871C.
March 1972	840-08946D	845-06894D	841-00619D 843-00406D 846-00444D 848-00376D	Increased front wheel castor angle.
April 1972				Six-point fixing replaced four-point type on left-hand exhaust manifold cover, to improve sealing for AED.
				Starter inhibitor and reverse light switch with automatic setting.
June 1972				Modified gear selector linkage to give more positive operation of inhibitor switch, from engine number 840-16746A.
September 1972				Plastic end added to wiper blades to prevent damage to decker panel.
				Improved distributor (Lucas LU 41393) with concentric base-plate in place of offset pivot, and with Zytel heel for contact points, from engine number 840-17926C.

1973 model-year

October 1972				Engine serial number now located on pad beside oil dipstick, rather than at rear of engine; engines now with suffix letter D.
				Lip type rear main bearing oil seal replaced rope-type seal.
				Constant-diameter hose between heater and engine; modified engine outlet pipe to suit.
April 1973	840-10158D	845-07945D		Modified end cover and oil seal for PAS boxes.
				Champion L92Y sparking plugs replaced L87Y type, to prevent misfires, from engine number 840-19940D.
June 1973				Final cars built.
March 1974				Manual choke conversion kit available, to counter supply difficulties with replacement AEDs.

SALOONS IN GOVERNMENT AND MILITARY SERVICE

During the middle sixties, the cars used for the transport of senior British Government officials were Humber Imperial saloons, invariably painted black. However, the Rootes Group dropped this limited-volume hand-finished car in 1967, leaving a small but significant gap in the market. The Jaguars of the time were considered far too sporting for the job of a gentleman's carriage; BMC planned to drop the unsuccessful Vanden Plas 4-litre R in 1968; and that left the Rover. It was an ideal choice for the job, not least because in 3.5-litre guise the car was new in 1967.

From the beginning of 1968, the 3.5-litre and 3½-litre Saloons thus became the trans-

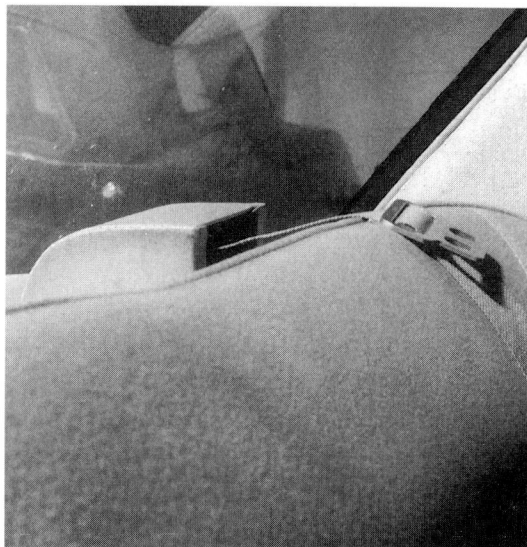

Many of the cars used by Her Majesty's Government went to Hooper's in London to be re-upholstered in cloth instead of leather, and to have inertia-reel rear seat belts fitted. The car pictured has both, and is actually the second one built for Prime Minister Harold Wilson in 1971.

British Military P5B Saloons

The nine batches of cars known to have been delivered to the War Department or Ministry of Defence were registered in the following sequences (batch totals are given in brackets):

64 FG 13 to 64 FG 19	(7)
20 FH 27	(1)
43 FH 46 to 43 FH 57	(12)
25 FJ 84 to 26 FJ 01	(18)
00 FL 60	(1)
00 FM 70 to 00 FM 77	(8)
01 FM 56	(1)
04 FM 55 to 04 FM 56	(2)
08 FM 34 to 08 FM 50	(17)

When sold onto the civilian market, they were of course re-registered. In some cases, cars may have retained a military identification plate, possibly screwed to the engine side of the bulkhead, and this will carry the original military registration number and details of the contract under which the car was ordered.

---

port of Cabinet Ministers. The first car was delivered to the Ministry of Public Works in January 1968, and may well have been the one intended for Prime Minister Harold Wilson. The last one remained in service until the early eighties, although the replacement of the Government fleet began in 1979 when Prime Minister Margaret Thatcher started to use an armour-plated Daimler saloon. After about October 1972, Saloons destined for Government use were 841-series cars (RHD export specification) and were delivered in the first instance to Hooper's, the famous London coachbuilders. Hooper's modified their interiors, adding beige cloth seat facings and inertia-reel rear-seat belts with the reels concealed under neat housings on either side of the parcels shelf. The seat belts, front and rear, were coloured brown to tone in with the rest of the interior.

At least two cars were specially built for Harold Wilson. The first car was registered JUU 570 and was probably 840-00151A. Like all the other 'official' P5B Saloons, it was finished in Ebony (black) with a White coachline. The car carried twin 12-volt batteries, one of which powered a radio telephone, and had some custom-built electrical equipment. Its interior was in Buckskin, and the car was delivered with both front and rear head restraints. It also had a unique ashtray in the centre console, designed to carry Wilson's pipe. This ashtray was designed in the Rover Styling Department by Tony Poole, and was based on the rear ashtray of a Coupé.

The second car was 840-09357D, and was built in September 1972. It was delivered to Hooper's for the usual interior modifications outlined above, but additional reading lights in the front and rear compartments were fitted at Solihull. The car once again had a radio telephone and twin 12-volt batteries, as well as additional electrical equipment which included twin ignition coils and various identifying marker lights. For this car, the special ashtray was fitted in a modified armrest in the left-hand rear door, and was accompanied by a small holder for a box of matches. The car was registered as PRK 315K and was bought

(Left) The special smoker's companion made for Harold Wilson's car appears to have been transferred from the Prime Minister's first P5B to his second. The ashtray is a modified Coupé type (from between the rear seats), and the indentation behind it is designed to carry a box of matches!

Even after Prime Minister Margaret Thatcher had switched to an armoured Daimler for official transport, Harold Wilson's P5B remained in use. Here it is outside No. 10 Downing Street in 1981, shortly before it was withdrawn from active service. It now belongs to the Heritage Collection.

More special features in the second Harold Wilson car: extra warning lights and switches have been fitted into Coupé instrument pods under the main binnacle. Note that there are no fewer than three key-operated switches: the main ignition switch is on the left of the steering column, but there is another key-switch in the earlier ignition-switch location at the bottom right of the main binnacle, and a third in the pod underneath it.

The console panel of Harold Wilson's second car has an array of additional switches. The one at the far left operates a reserve fuel pump, the switch next to it operates a siren, and to the right of that is one to switch on the light mounted just inside the windscreen which helped identify the car to Police. To the right of the cigarette lighter are switches for a spotlight, rear fog guard lamps and 'flash' – again to make the car stand out on the move if necessary. The radio-telephone was powered by a second battery installed in the boot.

by the Heritage Collection when it was withdrawn from service in the early eighties.

The example of the Cabinet was followed in other official circles. Black P5B Saloons were bought in quantity for ambassadorial transport overseas and for the transport of senior British military officials. Many had left-hand drive. These vehicles were maintained and operated by the War Department (later Ministry of Defence) on behalf of their users; in consequence, they carried military registration numbers. From about October 1972, the right-hand drive cars had 841-series (export specification) commission numbers, exactly like the cars intended for Government use.

Military records suggest that there were nine deliveries of P5B Saloons to the Ministry of Defence and its War Department predecessor, and that there were sixty-seven cars in all. However, there was probably at least one other delivery, as military records have not yet yielded end-user details of the final batch of twenty-one cars (841-00700 to 841-0711 and 841-00713 to 841-00721) which

The Royal P5B Saloons

Her Majesty the Queen had considered the Rover P5 her favourite car during the sixties, and when the 3-litre models were no longer available, she took delivery of her first P5B replacement. Rover records suggest that just two P5Bs were delivered for Her Majesty's personal use and that both were Saloons, although other P5Bs were undoubtedly used by the Royal Household generally.

The first of these two cars was 840-07057D, which was finished in a special dark green and delivered towards the end of January 1971. The Queen appears to have wanted to retain the registration number JGY 280 which had been on her earlier Rovers, but in fact the car was given the contemporary number JGY 280K – presumably because that was rather less conspicuous for a car which Her Majesty occasionally used to drive unaccompanied on public roads. That number was then transferred to a second car, 841-00723D, which was delivered in March 1974. This was also finished in a special colour, recorded at Rover as 'T&N Dark Green', and was the very last P5B to be made. The car is now in the Heritage Collection.

left the Rover Despatch Department for the military vehicle depot at Ashchurch in April and early May 1973. The first deliveries were made in 1969 and the last ones in 1973, although some cars were retained at Ashchurch until they were needed. This was probably the origin of stories current in the seventies that a large batch of Saloons had been 'stockpiled' for official use at the end of production. Withdrawal and subsequent sale onto the civilian market began in 1974 and continued until approximately 1982.

It is not clear exactly why the later right-hand drive 'official' P5B Saloons had export-specification commission numbers. A possible explanation is that these numbers helped to mark them out as different from the majority of standard-specification home-market Saloons and thus prevented confusion on the assembly lines. It is also possible that, as many of these cars were destined for service overseas, Rover found it simpler to build them all as export models.

Project Engineer Ken Stansbury remembers that a small number of these cars for the Government and military were armoured by a company in Wolverhampton. This company was probably MacNeillie, who later built

Gone, but not forgotten. Her Majesty the Queen's own P5B Saloon demonstrates where the inspiration for the new Rover grille came from in this 1991 publicity picture. The car in the foreground is the then-new Rover Sterling, a facelifted version of the 800-series which appeared in 1986 and was developed jointly with Honda in Japan. There was no engineering relationship whatsoever between the two cars.

armoured Land Rovers and Range Rovers. One armoured Saloon was built for the GOC (General Officer Commanding) Northern Ireland, some time after the British Army established a presence in that country during 1969, and others were retained in a Government pool in London for use mainly by Cabinet Ministers. Stansbury also believes that some armoured Saloons were built for overseas customers, quite possibly including senior Italian politicians.

SPECIAL ORDERS

For most of its life, the P5B was one of a very restricted number of low-volume, traditional luxury saloons on the market. As a result, many customers asked Rover for variations from the standard showroom specification.

The company was not prepared to go very far – there were never any manual transmissions or uprated engines, for example – but many more cars were delivered with special-order paint and trim than had been the case during the 3-litres' lifetime. Records show that several cars were painted in colours taken from the P6 options list, such as Mexico Brown (dark brown), Tobacco Leaf (light brown), Cameron Green (mid-green) and Almond (yellow). Ebony (black) was not listed for any Rover models, but it was surprisingly popular for both Saloons and Coupés.

The production records which have so far come to light do not reveal how many cars were fitted with special interior trim. However, several cars were certainly upholstered in beige cloth instead of leather, one of them being a 1969 Coupé which was painted in Tobacco Leaf over Mexico Brown.

The aftermarket or the afterlife? This 1971 Coupé was converted into a hearse by a firm of coach-builders for an undertaker in Sligo, Eire.

THE END OF THE LINE

There was no one single reason why P5B production came to an end in the summer of 1973. British Leyland rationalization has often been blamed as the primary cause, but the fact is that the basic design of the car was in any case fifteen years old by the early seventies and it could not go on for ever. Sales had begun to dwindle after the initial rush of enthusiasm for the new model at the end of the sixties, and it is clear that P5B production had always been intended to finish in the early seventies because the model's P8 (Rover 4000) replacement had been scheduled to enter production during 1971. If that car had not been cancelled because of the threat it presented to Jaguar's XJ6, P5B production might actually have ended earlier than it did.

There were other factors involved, too. Rover needed every V8 engine it could make on the single, overstretched production line at its Acocks Green factory. The V8-engined P6B (Rover 3500 and 3500S) models were selling strongly, and after 1970 the enormous success of the Range Rover further increased the demand for V8 engines. No doubt the production planners had begun to wonder about the wisdom of siphoning off a small number of precious V8 engines for a low-volume car like the P5B. And, of course, Rover could never find enough factory space. There was a new product waiting in the wings to use the space occupied by the P5B assembly line in the South Block and, as soon as that line had been dismantled in the summer and autumn of 1973, its place was taken by a new assembly line for the military 101in Forward Control Land Rover.

When a much-liked car ends its production run, the occasion is usually marked by a small ceremony. There appears to have been no such ceremony for the last P5B, a fact which has never been properly explained.

Probably, the fact that the last P5B was earmarked for Her Majesty the Queen had something to do with it. The Royal Household would not have liked attention being drawn to the car as it came off the lines, and would have asked for the discretion it had come to expect from Rover in these matters. Equally, the car was going to be specially finished away from the assembly lines, and would not actually be completed until several months after production had stopped. So, rather than have a ceremony centred on the second-from-last P5B, Rover chose to have no ceremony at all. In the rather demoralized atmosphere which prevailed at Solihull in the early seventies (during 1972, Rover had ceased to exist as a separate entity when its Engineering Department had been merged with Triumph's), few voices can have been raised in protest.

Some clues to the last days of production are nevertheless given by the records of the Rover Despatch Department, which are now held by the British Motor Industry Heritage Trust at Gaydon. These records show that production began to wind down in April, when the last right-hand drive export Coupés and Saloons were built, the majority of the latter being delivered to Ashchurch for military staff use. The final day was clearly 22 June 1973, when the last fifteen home market Saloons, the last left-hand drive Saloon, and the last home market Coupé were transferred from Production to Despatch. One car (840-10296D) was held back until 26 June, possibly because it needed some rectification work, but the Saloon destined for Her Majesty the Queen did not enter the Despatch Department until 13 July, three weeks after the assembly lines had closed down. Car number 841-00723D was not shipped to its new owner until 4 March 1974, by which time the P5Bs were no more than a memory in Rover showrooms.

9   Competition

It would certainly be no insult to the Rover P5 to describe it as an unlikely-looking competition car. In the seventies and eighties, it did become a favourite with banger racers, mainly because its strength enabled it to withstand plenty of the inevitable shunts which occur in that sport. However, during its production lifetime it was most definitely not regarded as a sporting machine. Even though the later V8-engined cars were quick for cars of their size, they were still basically big and heavy luxury barges, just as the original 3-litres had been.

Nevertheless, the 3-litre *did* enjoy a brief career as a competition machine, when it spearheaded Rover's entry into international long-distance rallying in 1962. During 1962, 1963 and 1964, three separate teams of P5 rally cars regularly surprised both onlookers and other competitors with their placings in some of the world's toughest rallies. There were no outright winners, and the 3-litres never achieved anything quite as glamorous as Roger Clark's sixth place overall behind the wheel of a Rover 2000 in the 1965 Monte Carlo Rally. Yet the 'works' P5s consistently demonstrated qualities of ruggedness, reliability and durability which spoke volumes for Rover's build quality. And while they emphasized the marque's traditional strengths, the very fact that the factory had entered them in any form of motorsport helped to create a new and more dynamic image of Rover as a marque. It was exactly what the Rover Board of Directors wanted to do before introducing the new and more obviously exciting 2000 in 1963.

The well-known story Rover's entry into motorsport was solely due to the efforts of its Managing Director William Martin-Hurst is in fact not true. Martin-Hurst was certainly an enthusiastic promoter of Rover's efforts in that direction, and he certainly was the man behind the development of the Rover-BRM gas turbine racing car of 1963. However, it was not Martin-Hurst who proposed that Rover should enter a team of 3-litres in international rallies, but rather Dan Clayton, who was P5 Project Engineer in the early sixties. As Clayton remembers:

> Rover was a very gentlemanly company at the time and I was looking at the P6 – which wasn't my project – and thinking that it was an entirely different animal. But nothing was happening to make any awareness in the public that Rover can be exciting. I thought, 'the image really surely needs to be changed. One way could be to use present models in a very restricted way to make people's concept of Rover change.'

By this time, Clayton had been to Kenya several times with 3-litres on proving tests (*see* Chapter 2), and he knew that the cars were immensely rugged, whatever their other faults.

> And I had the belief that we could enter a thing like the East African Safari Rally,

Competition

where it was toughness more than out-and-out speed that could achieve something. It was a chance of making people sit up and think, 'My goodness, Rover's got more than I realized'.

So Clayton put in a written proposal to enter a team of four 3-litres in the 1962 East African Safari Rally, routing it through his boss, Dick Oxley. 'I can almost say verbatim what my last paragraph was,' he remembered some 35 years later. 'I said, "The cost of this I estimate would be around £5,000 to enter four cars, which is the equivalent roughly of one half-page advert in the *Sunday Express*."'

The proposal was picked up almost immediately. Rover's senior management probably cast an eye in the direction of Humber, who had done their reputation a power of good with regular entries in long-distance rallies over the last decade, even though they had never again emulated Maurice Gatsonides' second place on the 1950 Monte Carlo event. The Super Snipe, which was in direct competition with the 3-litre in the marketplace, had been appearing in international rallies since 1959, and in 1961

achieved a very creditable fourth place overall in the very East African rally which Dan Clayton had suggested might also show the 3-litre to advantage. All of a sudden, Clayton found himself pushed into the role of Team Manager, with Engineering Shop Superintendent Ralph Nash as his deputy and the man in charge of preparing the cars. Neither man had any experience whatsoever of international motorsport, and nor was either of them particularly interested in rallying, so it was fortunate that there were a few fitters on Ralph Nash's staff who had at least had the experience of club racing in Britain! Other people were also drafted in from various other departments in Rover to help out on an *ad hoc* basis.

Towards the end of 1961, the Rover Company announced its intention to participate in a limited number of the 1962 season's international rallies. The three chosen were the East African Safari (in April 1962), the Liège–Sofia–Liège (in September) and the RAC Rally (in November). This announcement was met with a certain amount of scepticism at the time – not to mention incredulity – but the Rover team was well

On 23 January 1962, the four 3-litres which made up the first team of rally cars were lined up outside the main administrative block at Solihull for a photocall. On the left is Ralph Nash, Superintendent of the Engineering Workshops, who was responsible for their preparation. On the right is Dan Clayton, P5 Project Engineer and Team Manager.

How Rover saw the 1962 East African Safari

The 3-litres made their competition debut in the 1962 East African Safari Rally, and Rover management understandably took a keen interest in events. The following report was prepared for internal consumption in April or May 1962, apparently by the Rover Company's publicity officer, Gethin Bradley.

The performance of the Rover 3-litres in the 10th East African Safari made a considerable impression on the motoring public out there. Typical of the praise is a quote by Mr Eric Cecil, Chairman of the East African Safari Committee on the Kenya Broadcasting System shortly after the most difficult section of the Rally had been covered – 'Rovers are motoring so consistently that it almost seems as though they are using another drier, smoother road. Could you imagine a Rover 3-litre in any Safari, let alone this one? I couldn't, but by golly I can now. Even if they were to disappear – which won't happen naturally – even if they were to disappear into the Indian Ocean at Dar, nothing could detract from their amazing, even stately performance.'

Members of the press also took this point of view and two quotes follow – (1) By Gordon Wilkins of the *Observer* – 'Their performance obviously made a good impression on the East African crowds.' (2) By Philip Turner in *The Motor* – 'One of the sensations of the Safari was the outstanding performance of the team of Rover 3-litre cars not generally associated with rocketing through the African bush at very high speed; yet, at the end of the first stage of the rally a Rover driven by Ronnie Adams and Peter Riviere was leading the big-car class. This particular car was later one of many forced to retire, but two of the team of four cars finished the Safari.' These comments were taken up further by the East African Press and the coverage of the Rover team, both by press and radio, was extensive.

The Rover team of drivers had been instructed to drive steadily and to aim at finishing the course. If this meant losing time on a section, they should do so. With these tactics the cars arrived at Arusha in the second leg of the rally leading The Manufacturers' Team award, 1st and 2nd in the Big Car Class and 5th in General Classification. By this stage Ford Zodiac IIIs' 'Manufacturers' team had been put out of the running due to the fans cutting through their radiators, and two Mercedes were having suspension trouble.

Over the Mbulu Escarpment section – the most difficult – the small cars got through in the dry but the bigger cars were caught in rain and extremely difficult mud conditions. Here the Rover 3-litres excelled themselves. These conditions saw the final demise of the Zodiac and Mercedes team hopes. It is interesting to note that out of the six Zodiac Mk IIIs which started two finished, and that five Mercedes started and two finished. On the section following Mbulu the Ronnie Adams 3-litre had an accident when he ran into a wash-away. The Brochner car overtook Adams' position and did extremely well until Brochner was injured as a result of the jack slipping whilst he was underneath the car checking his sump shield.

He managed to carry on driving, but shortly before Mombassa he put a wheel over the parapet edge of a bridge. To clear the course the following Rally cars dragged the 3-litre off sideways and damaged the suspension beyond immediate repair. The remaining two 3-litres continued to finish. The Vincent family, having driven consistently throughout the Rally had no mechanical faults apart from the complaint that 'the cigar lighter failed to function', came

3rd in the Big Car Class. The Englebrecht-Goby car finished 5th in its class, again with no mechanical troubles apart from that caused by picking up a load of dirty fuel on route which cost them a considerable amount of time on many sections. All four cars were back in Nairobi in good mechanical condition within a few hours of the end of the Rally. None of the four Rovers needed any repairs or maintenance, other than the normal check-overs throughout the Rally. This is a very different story from that of the majority of other cars competing in this event.

Entry of these cars into the Safari and the news made by them in East Africa, together with Press stories circulating regarding the Rover T4 going to the States, and the Land Rover Vickers Hover Craft, made a tremendous impact in East Africa and generally gave the impression that the company was taking on a new burst of activity all of which appeared to be laying a foundation for future sales. Even if the results of the Rally do not take immediate effect on the sale of 3-litres, I consider that general Company Public Relations were excellent.

It is thought to be a good idea to bring the cars back to this country for inspection by the Rover Engineering Department. Perhaps useful propaganda could be promoted by sending one or two of them around to selected distributors.

Summary

104 cars started	– 45 finished
17 English cars finished	
6 Ford Zodiac Mk IIIs started	– 2 finished
5 Mercedes 220SEbs started	– 2 finished
3 Humber Super Snipes started	– none finished
4 Rovers started	– 2 finished

Rovers were 3rd in class, 5th in class, 25th overall and 31st overall.

Rover entries 4 cars:
1. Vincent and Vincent
2. Goby and Englebrecht
3. Adams and Riviere
4. Brochner and Gill

made in the name of Cooper Motor Corporation Ltd, Nairobi. Cars prepared by Engineering Department (Mr Oxley). Team managed by Mr Clayton and Mr Nash. Publicity cover: Mr E.H.G. Bradley.

organized by this stage. Four cars were taken from the assembly lines in January 1962 and, after initial preparation in Ralph Nash's workshops, were shipped out to Kenya where service support was to be provided by Rover's East African distributors, the Cooper Motor Corporation of Nairobi.

Rootes' 1961 success in the East African Safari with the Humber Super Snipes had been largely due to their use of local drivers, and Rover followed suit by engaging Gordon Goby and Per Brochner to drive two of the

Dust and unmade roads characterized the East African Safari Rally. Car number 103 was driven by Ronnie Adams and Peter Rivière on the 1962 Safari but crashed out of the event.

(Below) Car number 105 was crewed by Per Brochner and Sam Hermon-Gill on the 1962 East African Safari Rally, and was also eliminated after an accident.

team cars. Donald Vincent and his son Tony took the third car, and the fourth went to Briton Ronnie Adams, no stranger to large cars in rallying as he had driven the big Mk.VII Jaguars on the Monte Carlo Rally in the fifties.

As this was Rover's first competitions entry, the works drivers were instructed to do their best to get the cars to the finish in one piece rather than wreck them by attempting to win: Rover's main publicity aim was to show how durable the 3-litre was. In fact, the Ronnie Adams and Per Brochner cars were both eliminated by accidents, but

not before they had put up an astonishing performance which delighted the spectators and amazed the press. The Vincents brought their car home to a very creditable third place in class, and Gordon Goby's car finished fifth in class. It had been a remarkable start to a most unlikely venture.

THE 'DNX' TEAM CARS

The 3-litres' showing in the East African Safari might not have been spectacular, but it was extremely encouraging to Rover

Gordon Goby and Fairey Englebrecht had the bonnet and wing tops of their 3-litre painted matt black to prevent reflections on the 1962 Safari. Car number 101 came fifth in its class and 31st overall.

management, who agreed that the planned rally programme should go ahead with an entry in the Liège–Sofia–Liège Rally in September. For this, four more cars were specially prepared at Solihull, drawing on the lessons learned in the East African Safari. This time, the cars had the more powerful Weslake-head Mk II specification which was to be announced that autumn. They were registered as 676 to 679 DNX in July (or possibly August) 1962, and in that month Rover formally established its Competitions Department with Ralph Nash in charge. Dan Clayton bowed out and returned to his full-time job as P5 Project Engineer.

The rally regulations of the time insisted that the cars should be essentially to showroom standard, although a few modifications were permitted. To save weight, all the sound-deadening material was removed

This publicity picture taken in August 1962 purports to show two of the Rover team cars and their drivers. In fact, both are actually service cars, and both had probably been members of the first rally team. Nearest the building is Ralph Nash, by this time Team Manager.

677 DNX was a veteran of two other rallies when it was entered in the 1963 East African Safari, where it was crewed by Bill Bengry and Gordon Goby. The car came a highly creditable seventh in what the British press called 'a sensationally destructive rally.' Note the Rover nameplate on the front wings, fitted for publicity purposes.

from the bodies. Non-standard front seats were fitted, and the wiring was re-routed through the passenger compartment to make it more accessible in an emergency. Unlike modern rallying practice, however, the brake pipes remained outside and underneath: the braking system of the rally cars was completely standard except for the use of the hardest option production brake linings then available. The road springs were also the toughest production option – the so-called 'high suspension' springs designed for certain export territories – and non-standard dampers were fitted. Adjustable washer jets sited on the bonnet also anticipated their 1964 arrival on production cars by some years.

Like every other manufacturer, Rover interpreted the regulations about 'showroom standard' as liberally as possible. So, for example, when the Experimental Department was blueprinting the rally cars' engines on behalf of the Competitions Department, they examined dozens of valve-springs to find the stiffest production ones they could. Test engineer Brian Terry remembers that the ones eventually chosen were so stiff that they would almost certainly have been rejected from the production line as faulty! In

addition, the inlet ports in the cylinder head were polished to the maximum to give a power increase of around 10bhp; Brian Terry remembers that this was rather a tricky job and that a number of heads were ruined when the ports were polished right through to the water-gallery! The exhaust system used only a single silencer, which made for a power increase of around 8bhp and contributed a crisp exhaust note to the generally high level of noise inside the cars.

Nevertheless, Ralph Nash kept a watchful eye on the modifications his enthusiastic staff attempted to make. 'Ralph was Rover right through the middle,' smiles Lou Chaffey, who was a member of the service crews which supported the 3-litres after their first outing on the East African Safari.

We were always trying to push him a bit further but he only ever went as far as he dared. A Rover rally car still had to be a Rover. We could polish heads but we never messed around with different sizes of valves or anything like that. We built one P5 with aluminium panels and we lightened it by boring holes all over the place but it was never entered as a works car. I think it went to a privateer somewhere.

The DNX-registered cars made their first competitions appearance as planned in the Liège–Sofia–Liège Rally in September 1962. New drivers were engaged to replace the Kenyans who were naturally reluctant to travel half-way round the world for the European rallies which constituted the rest of Rover's 1962 programme, and this time out, the Rovers actually bettered their performance on the East African Safari. Although two cars were eliminated by accidents, the other two were among only eighteen cars out of the 100 starters to finish – and Ken James took sixth place overall.

The very high rate of attrition was a result of the demanding conditions in this 4,000-mile rally, as Bill Boddy of *Motor Sport* summarized in that magazine's October 1962 issue. 'This rally, over appalling roads,' he wrote, 'tested suspension and brakes as well as engines, and some terrible things happened. Even the winning Mercedes-Benz's brakes were a warm red hue after the crossing of the Moistrocca pass.' Boddy went on to point out that the 'remarkable Rover' won its class, and added that its overall sixth placing,

> with such an improbable rally component as power steering, was a fine debut for these more powerful new Rovers from the Birmingham factory. This second-highest placing by a British car in Europe's most gruelling rally [*the highest was an Austin-Healey, in fifth place*] is a fine breakthrough for the Rover Company's desire to kill the impression that it manufactures only sedate old-men's cars.

The Rovers' next competition appearance was in November on the RAC Rally, a 2,200-mile chase through fifty counties which started and finished in Blackpool. A feature of the event were the timed special stages, held over 300 miles of Forestry Commission roads, but the Rovers did not excel on any of these and their best result was Johnny Cuff's sixth-fastest time on the first of the Bovington Camp stages. Nevertheless, Ken James managed to bring his 3-litre home to eleventh place overall and despite a lack of class wins, the Rover team were more than pleased with third place behind BMC's Minis and Triumph's TR4s in the Manufacturers' Team Award.

Rover, was of course, still new to the competitions business, although the company was learning fast. Lou Chaffey remembers:

> It was very amateurish at first. I remember one RAC Rally when we were told to find the service point in Darlington. They didn't tell us where it was but they thought it would be pretty difficult to miss it. So off we went, two of us in this service car, a day early and in plenty of time. And we spent hours driving up and down all the main roads in Darlington and we couldn't find it. So eventually, we asked a policeman. He told us it was nowhere near where we were, so we tore off at full speed and only just got there in time. We found out that the next service point was in north Wales and I'd had enough by this time, so I stopped and bought some Ordnance Survey maps to help us find it. Can you imagine that sort of thing happening these days?

By the time of the DNX-registered 3-litres' final competition appearance, in the 1963 East African Safari, the Competitions Department had moved from a corner of Ralph Nash's workshops to a screened-off area within the newly-built North Block, which had already started building Rover 2000s in anticipation of their introduction in October 1963. For this year's Safari, the Kenyan drivers Per Brochner and Gordon Goby were again persuaded to drive for Rover, although Goby this year partnered regular driver Bill Bengry. Regulars Ken

Flowers for car number 21 – the 3-litre of Ken James and Mike Hughes came sixth overall in the 1962 Liège–Sofia–Liège Rally which was the first outing for the DNX-registered cars.

James and Johnny Cuff drove the other two 3-litres.

However, the 1963 Safari was, as Bill Boddy wrote in the May 1963 *Motor Sport*, 'a sensationally destructive rally'. Car after car was eliminated by accidents, and of the eighty-four which started only seven reached the finish. Peugeots came first, fifth and sixth, and as Bill Boddy commented:

High praise must be accorded to the four other finishers – to the British Ford Anglia 1200 which was second, to the Mercedes-Benz 220SEb that was third, to the Fiat 2300 which came in fourth and to the 3-litre Rover P5 which was seventh. East Africans and all the World's customers will long remember that for toughness and reliability, you want Peugeot, or Ford Anglia, Mercedes-Benz, a six-cylinder Fiat or one of the new big Rovers!

The seventh-place car was Bill Bengry's 3-litre – and Rover quickly seized on its performance to make a publicity point. An advertisement in *Autocar* for 26 April 1963 proudly announced, 'Rover 3-litre Saloon wins Big Car Class', and went on to explain that,

not only was Mr Bengry's Rover the only finisher in the over 2,500cc class (without any mechanical trouble whatsoever!), but he had the distinction of being the only UK driver to finish. This particular car, a perfectly standard 3-litre Saloon with extras available to all Rover owners, competed successfully in last year's Liège–Sofia–Liège and RAC Rallies – a hard life even for one of the world's best engineered cars.

It is, of course, arguable whether all the extras on the car would have been available to Rover owners, and it was unfortunate that the car pictured in the advertisement was actually Bill Bengry's Safari mount of the previous year, and not his 1963 car at all!

THE 'GNX' TEAM CARS

The DNX-registered cars were retired from front-line duty after the East African Safari and a new team of four 3-litres took over.

However, the older cars' working lives were far from finished, as they were now designated for practice and as the transport for the service crews. Lou Chaffey remembers that some of them were quite heavily modified and were very probably quicker than the actual team cars. He points out,

> they had to be because we were carrying tons of spares – the all-up weight of a service car was about three tons – and we had to get to the service points before they did. There was great friendship among all the service crews, and we got on well with BMC and all those people. We used to race each other to the controls. You'd get a Rover 3-litre with a great big wooden box of spares on the roof and we'd be hanging the tail out round corners and taking the straights at 100mph or more. It's frightening to think about it now, because they didn't close the roads in those days and normal life carried on while the rally blasted through. It seems totally irresponsible now.

The new team of 'works' 3-litres took to the roads early in June 1963, after spending some three months being prepared by the Competitions Department. Brian Terry remembers that they were fitted with the 'big crank' engine, which was still under development and would not actually appear on production cars until the autumn of 1964. All of them were finished in Pine Green – the nearest standard Rover colour to patriotic British Racing Green – and they wore the registration numbers 996 to 999 GNX.

The GNX-registered cars had their first outing on the 1963 Spa–Sofia–Liège Rally (a re-routed and renamed Liège–Sofia–Liège), driven by regulars Bill Bengry, Ken James and Johnny Cuff, and by a new recruit to the Rover team, John Sprinzel. Always a tough rally, the 1963 Liège was described by the top Swedish rally driver Eric Carlsson as the toughest yet. Its 3,430 miles eliminated all but twenty of the 120 cars which entered – and among those twenty survivors were two of the Rover 'works' 3-litres. Ken James put in the best performance, finishing eighth overall and winning the over-2,500cc class, and his performance did the Rovers credit. As Bill Boddy put it in the October 1963 *Motor Sport*, 'any car that gets home in the Liège isn't to be

Examining the damage at a service stop at Gorizia on the 1963 Spa–Sofia–Liège Rally. The 3-litre behind, carrying two spare wheels in its boot, is one of the service cars.

*Ken James and Mike Hughes crewed 999
GNX to eighth place overall on its first
outing in the 1963 Spa–Sofia–Liège Rally.*

despised'. Nevertheless, it was by this stage becoming increasingly obvious that the 3-litres were never going to finish in the top handful of cars in the overall classification of an international rally. The new 2000 was also just about to be launched, and Rover had every intention of preparing that car to do battle in tough international rallies as well. In due course, Rover 2000s would replace the 3-litres in the 'works' team altogether – but the rally world had not seen the last of the P5s yet.

Five 3-litres competed under the works banner in the RAC Rally that November, four of them being the current team cars and the fifth being one of the DNX-registered cars which had been privately prepared by Toney Cox. For this event, Rover had come up with a new publicity gimmick to emphasize the reliability of the P5s: each car would carry its own spares, and there would be no service crews so that the drivers would have to do their own repairs during the rally. Richard Martin-Hurst, son of

996 GNX was one of the third team of cars, and was driven in the 1963 RAC Rally by John Sprinzel.

The ex-works team cars continued to appear in major events after their careers with Rover were over. This is 678 DNX at Devil's Bridge on the 1964 RAC Rally, when it was driven by Major Freddie Preston and Staff Sergeant Roy Davies of the British Army Motoring Association. This kind of damage on such events was expected.

Rover's Managing Director, was paired with Roger Clark – who would drive a Rover 2000 to sixth place overall in the 1965 Monte Carlo and became a noted rally champion – but they allegedly had a difference of opinion about driving styles and Clark stayed in the navigator's seat throughout! All five cars finished the event, although none of them shone on the special stages. Volvo beat Rover to the class win while Ford and Volkswagen were ahead in the manufacturers' team prize, but the result provided yet more valuable publicity.

Rover probably planned to use the new 2000s for all the 1964 season rallies, but the cars were not ready in time and so the GNX-registered P5s were pressed into service once again. All four ran in the Acropolis Rally in Greece during May, this time with yet another set of drivers. Only Ken James remained from the 1963 'works' team, and the other three drivers were Anne Hall, Logan Morrison and Toney Cox. This was a particularly tough Acropolis Rally, and its results were decided largely on the cars' durability and on the results of the twelve groups of special stages and the three special hill-climbs. *Motor Sport*'s correspondent reported in the magazine's July 1964 issue that,

> even the normally reliable Rovers had their troubles and their fastest car, that of Logan Morrison, retired just after the Amphissa hill-climb when a rear wheel fell off. The highest placed Rover, driven by Ken James and Mike Hughes, did very well to finish sixth despite having trouble with their brakes.

And that sixth place overall was indeed creditable, for the Rovers had been beaten only by a pair of Mercedes 220SE models, Pat Moss' Saab, a Citroën DS19 and the winning Volvo PV 544 of Tom Trana.

However, the Acropolis was to be the 3-litres' last success in the 'works' rally team. A team of Rover 2000s represented the Rover factory in the Alpine Rally, and the 3-litres were entered only once more in the team's name. This was in the Spa–Sofia–Liège event in September 1964 when two 3-litre 'works' entries ran alongside two 'works' Rover 2000s. Two private entries in fact made the number of 3-litres in this rally up to the familiar four, but none of them enjoyed much success. *Motor Sport* reported in its October 1964 issue that

> The slogan of Rover reliability was not aided by the fact that none of the four 3-litre Rovers entered in the rally finished, with crashes eliminating two of them and a punctured radiator and faulty distributor taking care of the other two.

So on that rather downbeat note, the 3-litres were retired from the Rover rally team, and the 2000s took over for good. Nevertheless, the P5s continued in use with the Competitions Department as service cars until the department was closed in 1966, and some of them were borrowed from time to time by privateers or by the British Army Motoring Association, who entered them in some of the lesser rallies. By the middle of 1967, however, the cars' careers were over, and Rover scrapped those which remained in its ownership. Just two may have escaped the crusher: 676 DNX, which had been sold to Richard Martin-Hurst in 1963, and 999 GNX, which was sold to Cooper Motors of Nairobi some three years later. Their subsequent fate is unknown.

Reflecting on the recent Spa–Sofia–Liège Rally in its October 1964 issue, *Motor Sport* noted that: 'Whenever anyone asks exactly

The final appearance by the 3-litres under the factory banner was in the 1964 Spa–Sofia–Liège rally. 997 GNX was of course one of the third team of rallying 3-litres; a Rover badge was once again fitted to the front wings for publicity purposes.

what do rallies prove, the answer is some variation on the theme of testing strength and reliability and then feeding the lessons learnt back into the production line. ' The rallying 3-litres had certainly helped to improve the breed: Brian Terry remembers that among the problems they revealed were weaknesses in the P5's damper mountings, gearbox bearings and engine pushrods, and that these items were all strengthened on later production cars. And Lou Chaffey confirms this view as he fondly recalls his days with the Rover service crews:

> We had a lot of fun and the public got a lot of benefit, too. The production cars were changed in many ways as a result of our experiences, and I think Rover got its money's worth out of the programme.

The 'works' rally 3-litres

There were three teams of four 'works' 3-litre cars, making twelve cars in all. The first four cars were Mk IA models, and the subsequent eight all Mk IIs. All twelve cars were Saloons.

The first team
The identities of the first four team cars have not been established beyond doubt. They were Mk IA models, built towards the end of 1961 or in January 1962. Two of them appear to have been finished in Ivory, and the other two in a dark colour. The cars were prepared initially at Solihull and final preparation was carried out with the assistance of Cooper Motors of Nairobi, who also registered them in Kenya.

These first four team cars appeared in the East African Safari Rally in April 1962 and were then brought back to Britain. Two were probably re-registered as 674 and 675 DNX, and became service cars on the 1962 Liège Rally. 675 DNX was 725–00335A and was sold to Toney Cox in July 1963. The eventual fate of all four cars is unknown.

The second team
The second team of four cars was specially built at Solihull to Mk II specification. The cars were registered in July or August 1962, and appeared in the 1962 Liège–Sofia–Liège rally and the 1962 RAC Rally. They were then relegated to practice and support duties, and were used by the Rover service crews on several subsequent rallies.

The cars were:
770-00101ECS which was registered as 676 DNX and had two-tone paintwork
770-00102ECS which was registered as 677 DNX and painted Burgundy over Stone Grey
770-00103ECS which was registered as 678 DNX and had two-tone paintwork
770-00104ECS which was registered as 679 DNX and had two-tone paintwork, possibly Stone Grey over Burgundy.

The ECS suffix probably stood for 'Experimental Competition Specification', and the serial numbers were not taken from the normal Mk II Saloon sequence; 770-00101A to 770-00104A were all ordinary home market Saloons. The four rally cars all had matching engine numbers, e.g. 770-00101ECS had engine number 770-00101ECS. Engine number 770–00105 was fitted to 675 DNX, one of the first team cars.

676 DNX was sold to William Martin-Hurst's son, Richard, in October 1963 and was used by him in a number of rallies. 677 DNX was scrapped in February 1964. 678 DNX was used by a BAMA (British Army Motoring Association) team in the 1965 Scottish Rally, where it won its class and came 12th overall. The car was scrapped in February 1967. 679 DNX was probably the car which a second BAMA team (Major Freddie Preston and Staff Sergeant R. Davies) drove to second place in their class on the 1965 Scottish Rally. It was scrapped in January 1967.

The third team
The four cars which made up the third Rover 'works' team were built in February and March 1963, and were all registered at the same time on 7th June. All of them were painted in Pine Green.

The cars were:
770-00769B   registered as 996 GNX
770-00768B   registered as 997 GNX
770-00770B   registered as 998 GNX
770-00758B   registered as 999 GNX

996 to 998 GNX were all scrapped in April 1967. 999 GNX was sold to Cooper Motors of Nairobi in May 1966, and its subsequent fate is unknown.

10   The P5 and P5B Today

The last P5Bs were built nearly a quarter of a century before this book was published, and relatively few remain in everyday use even though the cars survive in reasonably large numbers. Like many other cars of their period, they seem large, cumbersome and slow by modern standards, and their thirst for fuel discourages most owners from using them as regular transport.

It was this thirst which was largely responsible for the very rapid demise of the 3-litres, which are now surprisingly rare by comparison with their younger sisters. The newest of the 3-litres was just six-years old and the oldest was around fifteen-years old when the 1973 Oil Crisis caused petrol shortages and was followed by a dramatic increase in pump prices. Cars with heavy fuel consumption became a liability almost overnight, with the result that owners of older 3-litres were unwilling to spend money on cars which became in need of major repairs. So it was that many of the older examples found their way to scrapyards, and a large number of the newer ones suffered the same fate after the second Oil Crisis in 1979. The V8-powered examples – newer and rather more glamorous – were rather less affected by such problems, although prices dropped and many cars fell on hard times and were sadly abused.

Fortunately, the eighties saw a revival of interest. As so often happens in the classic car world, it was the models with the highest performance and highest levels of equipment which first attracted enthusiasts, and

the V8-powered Coupés quickly attained a pre-eminence which they have never lost. The Saloons were not long in following, however, and now at long last the excellence of the 3-litres is being appreciated by the classic car movement. It is sad that the survivors – particularly of the very early cars – are so few in number.

Two very welcome results of this revival of interest have been the establishment of owners' clubs dedicated to the P5 and P5B and the rise of parts and restoration specialists who cater for them. There was a time in the late seventies and early eighties when running a 3-litre could sometimes seem a rather lonely experience, and when the only sources of spares seemed to be uncaring British Leyland dealers or scrapyards. Things are different today, however, and while the sheer cost of remanufacturing some items means that it can be difficult to find some parts, most of the consumables needed to keep a car in running order can be located by little more than a telephone call.

Even so, the P5s and P5Bs are not cheap cars to own. They were built to very high standards of engineering and were designed to be run by the wealthy professionals of their day, so it should not be surprising to discover that repairs and running costs can be quite expensive. This has a fairly predictable effect on market values. Cars which are in really poor condition simply cannot be given away, because restoration costs would be prohibitive. Cars in first-class original

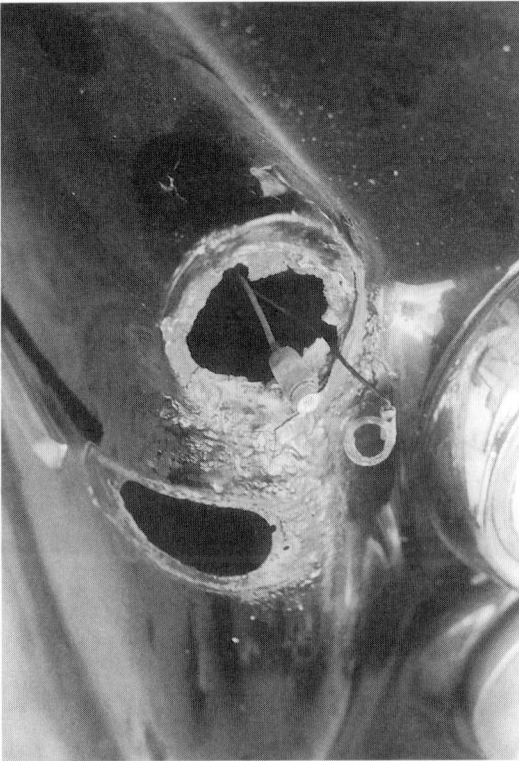

Rot around the sidelights is a common problem on P5s of all ages.

CHECKING FOR ORIGINALITY

When looking at a P5 or P5B for sale, it is generally easy to tell whether the car is in genuinely original condition or not. The factory-applied paintwork was of a very high standard, and Rover's inspectors would not have tolerated the orange-peel finish which is accepted as normal on many modern cars. Panel gaps were invariably even and consistent, and it is often easy to spot a replacement front wing on a P5B because many of those made in later years are a poor fit around the scuttle panel and will not line up properly with the edge of the bonnet. The problem was that the tooling had worn and, under British Leyland, nobody cared enough to have it put right after the cars had gone out of production.

On P5B models, the most obvious giveaway of replacement lower panels or a partial respray is likely to be the coachline. Rover never did provide detailed instructions of how to reconstitute the coachline (which was painted on by hand, not stuck on as so often happens with modern cars), and so repairers and restorers had to do their best. They usually got it wrong. Coachlines should run parallel to the bright trim strip and an even 30mm away from it. They should be the right colour for the car (*see* the Table in Chapter 8), and their ends should be correctly shaped (see the illustrations which accompany this chapter). Saloons should not have a coachline on the roof, of course, as this feature was reserved for Coupés only.

Seats and door trims tend not to be much of a problem on allegedly original cars, but carpets and wood trim need careful examination. Carpets should be a first-class fit and should be held down by grey-painted spring clips, while the front passenger carpet should not have a heel-mat to match the one on the driver's side. Some replacement carpets have been a poor fit, particularly

condition, on the other hand, command sometimes astonishingly high prices because buyers know how much they would have to spend on restoring another car to get anywhere near that condition – and that the car would even then not be in genuine original condition. The majority of cars fall somewhere between the two extremes, though usually towards the bottom end because all P5s and P5Bs reveal expensive hidden problems when restoration work begins on them, and buyers prefer to allow for these unexpected costs. On the whole, market prices are therefore surprisingly low: not for nothing did *Classic and Sportscar* magazine's November 1988 issue describe the P5B as 'Britain's best-value classic car'.

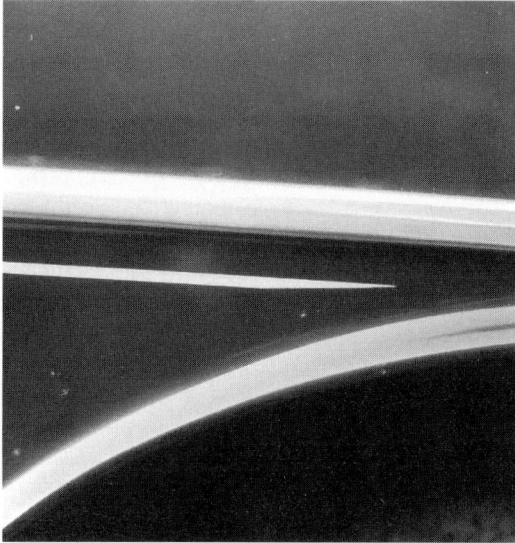

Coachlines on the V8-engined models are often 'restored' wrongly. They should terminate behind the front wheel arch like this ...

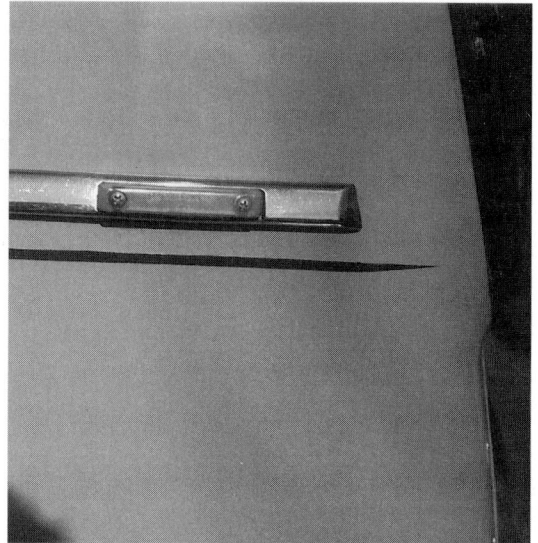

... and on the rear wing like this. (The slightly wobbly upper edge here was caused by masking designed to preserve the original coachline when the wing was resprayed.)

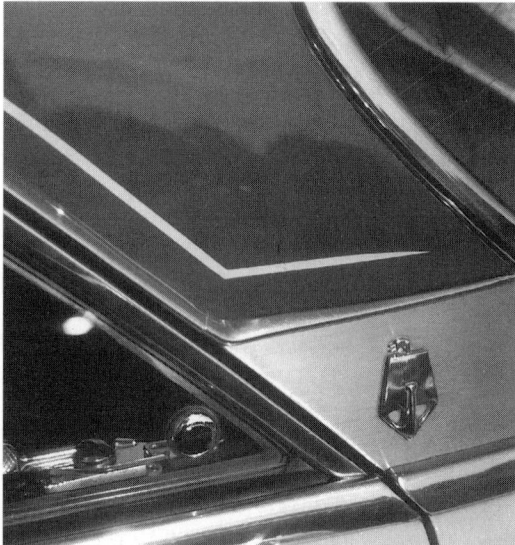

The coachline on the roof of V8 Coupés should terminate on the rear pillar like this ...

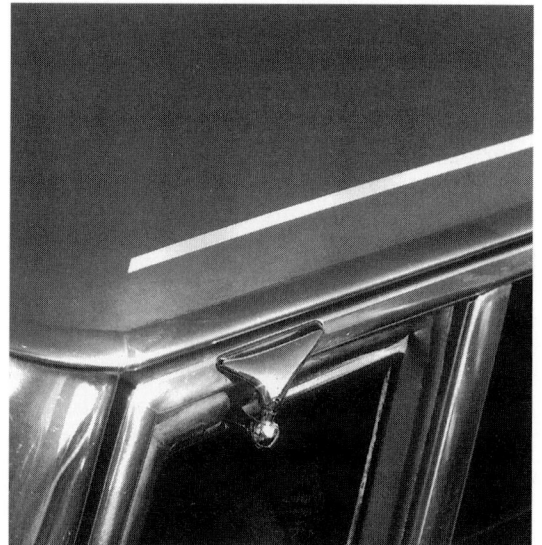

.... and at the front of the roof it should look like this.

over the transmission tunnel (and the moulded parcels tray of the 3-litre gearbox cover carpet has never yet been successfully reproduced), while there have also been some in rather dubious shades which would never have been allowed near a Rover by the car's makers. The wood trim appears to have been stained by hand using stockinette pads and then sprayed with a shellac varnish; this unique finish peels off and is extremely difficult to replicate. Suspiciously light-coloured wood should normally be considered non-original.

There are, of course many items to examine when checking that a car's detail specification is original. Wheel trims, radios, accessories and sometimes more major items have often been replaced down the years, and even well-intentioned owners have sometimes had to settle for an incorrect specification simply because a replacement original was not available. The tables and pictures in Chapter 2 to Chapter 8 should help to establish the 'correct' specification for any given car. Also worth remembering is that the British Motor Industry Heritage Trust has the despatch records for all P5 and P5B models, and that its Production Trace Service (telephone 01926-641188) can establish from these the date a car was built, its original colour and sometimes its original trim colour, and even the name of the dealer or distributor to which it was originally supplied.

WHAT TO LOOK FOR

Panels and Structure

Most cars offered for sale will not be in pristine original condition or anything like it. They are likely to be suffering from one or many of a whole series of well-known problems, and before examining a potential purchase it is sensible to know what these problems are and how to detect them.

The most common problem with all P5s and P5Bs is rust damage to the panels and to the monocoque structure, and this is often the most expensive to deal with. There are relatively few unrestored cars around which do not need repairs to the body sills, to the lower D-posts, and to the A-posts behind and above the front wheelarch splash panel. Complete replacement of all the affected areas is feasible, because remanufactured panels are readily available, but the welding involved needs skill, care and patience. To have a professional restorer do the job properly can easily cost as much as the purchase price of the car, and those who have never worked on P5s before often make mistakes which range from the irritating to the critical.

The outer skin panels are easy enough to examine and – because they all bolt on to the monocoque with varying degrees of difficulty – are easy enough to replace or repair in order to make a tired car look more respectable. For that reason, to assume that the monocoque underneath sound and shiny panels is rust-free would be to make a major mistake. Nevertheless, as far as the skin panels themselves are concerned, there are certain areas which need proper examination.

Starting at the front, the valance panel beneath the bumper of P5Bs often rusts badly, although its equivalent on 3-litre models usually suffers rather less. Front wings rot out around the sidelamp peaks and around the indicators, mainly because the area behind them traps mud and water against poorly-protected metal. A weld seam running diagonally downwards just ahead of the front wheelarch can also rust from the inside outwards. Then the lower trailing edge of the wing may disintegrate, taking with it the bottom mounting bracket. Lastly, rust may break out around the clips which hold the stainless steel trim in place and, on P5Bs, around the holes for the indicator repeater lamps and plate badges. Patch panels can be bought; full replacement wings sell for premium prices.

Another problem area is the outer edges of the scuttle panel, at the base of the windscreen. This is a double-skinned panel, and a leaking windscreen seal will allow water to collect between its two skins. The water generally runs to the outer edges and collects there, eventually causing the panel to rust through from the inside. Remanufactured outer sections can be bought, but it takes a skilled bodyshop to fit them properly and the amount of preparatory dismantling makes the job an expensive one. Further back along the car, the door bottoms often rust out when blocked drain holes allow water to collect inside. Even if the damage is not extensive, doors on the kerb side of the car often start to rust on their bottom edges where these have been scraped against kerbs while being opened. Replacement door skin lower halves can also be bought, but it takes skill to make a good job of butting these up against the top half of the original one-piece skin.

Below the doors run the outer sill panels, which are structural as well as cosmetic and therefore deserve very careful examination (see below). From the cosmetic point of view, however, any defects will be immediately obvious. On P5B models, a stainless-steel strip should be clipped to the outer sill, and this is often loose (because of broken clips) or missing altogether. Rust may take a hold around the clips, which are pop-riveted to the outer sill panel. Also worth noting is that the sill panel on all P5B models should be painted in black polypropylene paint (as should the bottom of the front wing panel on a line with the sill); restorers frequently forget to do this or are unsure about the correct specification. All four jacking points should have round rubber bungs with a deep circular indentation in the middle, and it is worth knowing that there are two types of these: the later ones have a larger diameter shaft and will not fit to early cars.

At the back of the car, the rear wings often rust at their leading edges after mud has built up behind the metal. Mud also gets trapped inside the wing trailing edge below

The rear valance panels behind the rear wheels rot through.

the bumper, while road-dirt thrown up by the wheels onto the outside is rarely cleaned off thoroughly. Between them, the two conspire to set up corrosion. Electrolytic action between the steel panels and the aluminium beading which fills the join between wing and body can cause the beading to corrode into a crumbly white powder. This is usually most apparent when the wing is off, but corrosion can also work through to the exposed top surface and cause the paint on the beading to start bubbling. Lastly, the two panels below the bumper on either side of the centre valance at the rear are notorious rust-traps and are almost always rust-damaged.

The P5s and P5Bs both carry a lot of bright trim, either in the form of stainless steel or chrome-plated metal of one sort or another. The stainless-steel items are rarely a problem (although side trim strips can bend), but the chrome-plated parts become pitted and then begin to corrode. Surface corrosion can usually be kept at bay by regular polishing, but once it becomes more deep-seated will become more expensive to deal with. The biggest headache for both P5 and P5B owners is the rear bumper, which is a massive one-piece item. It is vulnerable to parking knocks as well as to corrosion, and is particularly likely to rot through at its outer corners as the result of mud building up on the inside. New bumpers have been unavailable for some time, and although old ones can often be reclaimed by straightening, metal filling and rechroming, these processes are expensive. The cost of a reconditioned rear bumper is enough to discourage many people from buying an otherwise sound and cheap car. Fortunately, these expensive items are not subject to theft: most of them are practically welded to the car through corrosion of the mounting bolts!

Moving on to the inner structure of the monocoque, there are several areas to inspect. The splash-panels behind the front

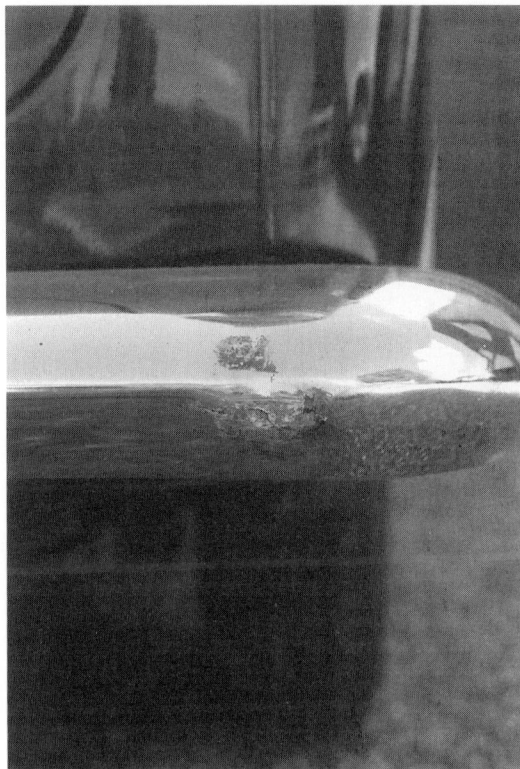

Rear bumpers tend to corrode at their corners, as shown here.

wheels are often rusted out, but they are reasonably easy to replace with remanufactured panels and do not contribute to the body's torsional strength in any case. However, the complex area above and behind them, where the front inner wing joins the bulkhead, also rots through, and this is a load-bearing area. Repairs are not easy for the amateur and are expensive when professionally done: many cars have been hastily patched to keep them roadworthy. Even the bulkhead can rust through in the area above the footwells, and there is often rust in the footwells, the front floor panels, and the body cross-member to which the subframe is mounted.

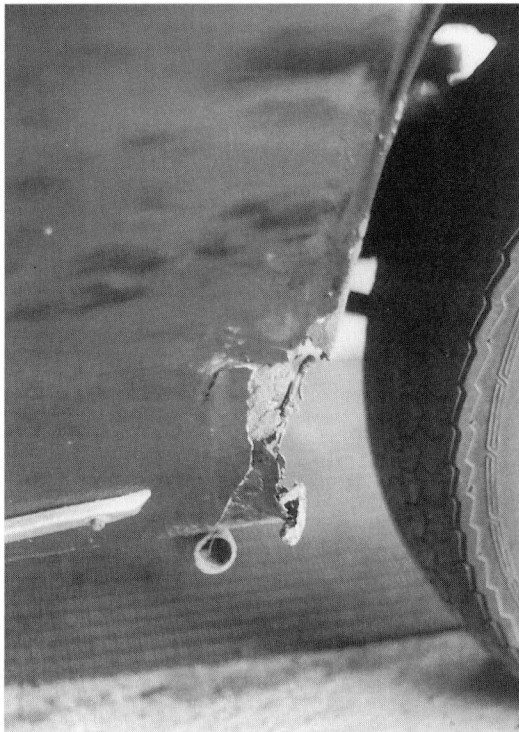

The bottoms of the front wings are another common problem area. This one has been poorly repaired and has rotted out again.

Behind that, the lower A-post may be weakened by rust, and in bad cases can actually crack under the weight of the heavy doors. The inner footwell side panels may also be rusty, either through contamination from the A-post or because of a long-term water leak from the windscreen seal. Repair sections for the A-pillar are available, but it takes skill to weld them into position correctly and invisibly. Rust damage in the A-post area can of course also contaminate the front of the sill and the front jacking point.

The sills, in fact, consist of three pieces, with the shaped outer section being welded to a vertical spacer panel and to the heavier main or inner sill which also bears the jacking points. Saloons and Coupés use the same three panels, although metal has to be cut from the outer sill to fit around the larger B-pillar on Coupés. Rust usually gets a hold on the inner sill first, and holes this so that water can get in. This water then sets up corrosion inside the sill, which goes on eating away at the metal even after the initial damage has been repaired. The only way to inhibit this corrosion is to pump the sill full of a rust inhibitor such as Waxoyl after drilling holes in its top surface (which should then be plugged with rubber bungs and will be concealed under the kick-plates). Most such anti-corrosion agents are inflammable, and it is worth checking that sills have not already been treated before starting any cutting or welding work; fire can break out only too easily and will spread very quickly, especially if the interior trim is still in the car.

At the back of each sill is the D-post, whose outer section is visible when the rear doors are opened as the forward edge of the rear wheelarch. These rust badly – ably assisted by water held against them in lower door seals which have turned into absorbent sponges over the years – and serious corrosion here will weaken the rear of the car's structure. The problem is not confined to the visible section, either. The D-post actually extends upwards behind the leading edge of the rear wing, and rust eventually makes its way unseen all the way up the D-post and into the outer wing panel and the inner wing as well. Remanufactured D-posts are available, but they are quite expensive and need to be fitted and welded in place by someone who knows what he (or she) is doing.

There are two other main areas of weakness at the back of the car. The first and less serious is the inner wings, which often rust right through. The second is the underbody reinforcing 'legs' which look rather like chassis rails and run under the boot area. These

(Above) *The bottoms of the doors suffer badly as well. This is the rear door on a P5B Coupé. The rubber door seal has become detached from its channel and is hanging loose, while rust has started to creep up the curved section ahead of the wheel. The sill on this car has also been rather crudely repaired, and both the rubber bung for the jacking point and the centre cap of the Rostyle wheel are missing.*

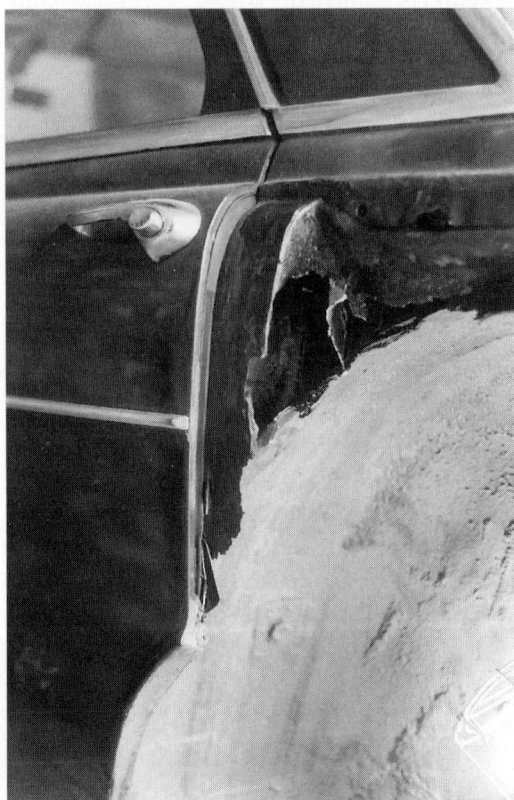

Rust damage to the inner structure of a P5 can be extensive, and will not be visible from the outside. The only indication of this problem on a 3-litre was a line of rust bubbles at the leading edge of the rear wing. With the wing removed, it became clear that the upper D-post to which the wing bolts was in an advanced state of decay.

can corrode badly, particularly above and behind the axle, and in serious cases will threaten the integrity of the spring mountings. Repair panels are available for both areas, but proper repairs around the area of the spring mountings are best left to professionals and will therefore not be cheap.

In theory, the front subframe can be detached from the monocoque to make major overhauls easier, but anyone who has attempted to remove a subframe which has been bolted in position for three decades or more will know that theory and practice are not at all the same thing! The subframe does have some weaknesses, often corroding in the box-section cross-member at its rear, and sometimes being weakened by rust around the steering box mounting. Repairs are always awkward with the subframe still attached to the car. Worth noting is that 3-litre and V8 subframes differ and are not interchangeable. The simple way of identifying a V8 subframe is by the two tubular members running from front to rear, which are absent from the 3-litre type. The engine mountings on the two types of subframe are also completely different, and of course very early 3-litres have a different set of engine mountings yet again.

Interior Trim

The first thing to remember is that a full retrim in high-quality leather to match the original will be enormously expensive. Tired or dry leather can be rejuvenated using hide food over a period of time, and discoloured leather can be successfully recoloured either by professionals or by using a DIY kit. Torn or damaged panels

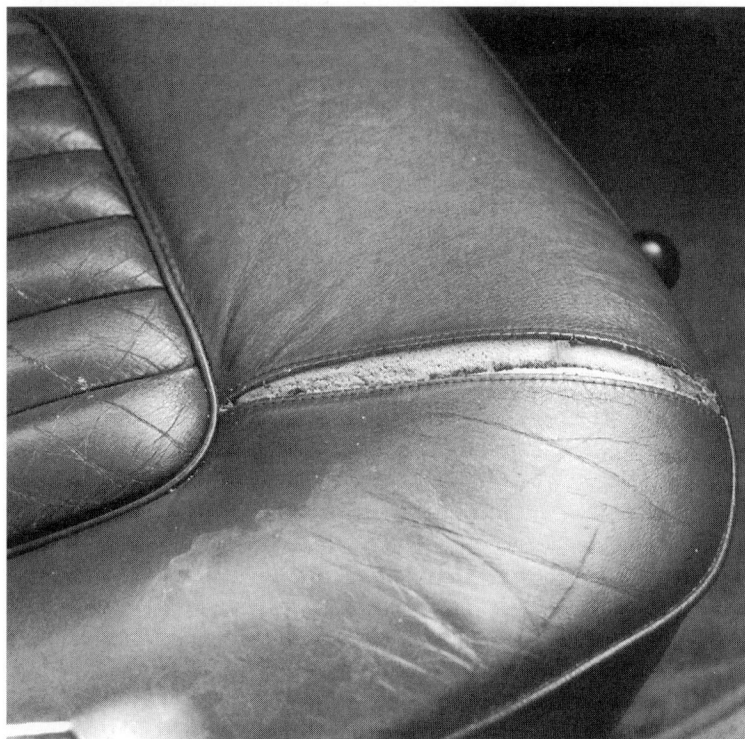

These seat seams are a weakness on Mk III 3-litres and V8-engined cars; the thread rots and the seam then opens up.

can usually be replaced individually by professionals.

There are of course three major types of seat associated with the P5 and P5B models, plus a few sub-variants. The first type was used on the very first cars and has pleats which run from side to side. It has no particular weaknesses, and nor does the second type, used between 1961 and 1966, which has pleats running from front to back. However, the top of the rear seat back on both types can discolour, dry out and split in the sunlight. The third type of seat, used on Mk III 3-litres and P5Bs, has stitched seams on the forward corners of the cushions, and these regularly split. They can of course be resewn, but the seat cover has to be removed from the seat carcass before this can be done satisfactorily; attempts to do it with the cover on the seat can provide a short-term solution but usually result in torn leather as well as broken stitches.

Early door trims with the charcoal-coloured contrasting panels do not wear well, and the leathercloth tends to come away from its backing panel. On the later type of trims, the stitches which hold trim to panel may come undone. The elastic which holds the map pocket to the main trim panel may also stretch or break, and the pile covering of the pocket itself may wear badly. The stitching around the top of the armrests may also give way, and the tops of the armrests themselves may split or wear into holes. The door trims attach to the doors by spring clips, which can tear away from the backing panel if handled too roughly or if the panel itself has been damaged by water. Remanufactured versions of the later trim panels can be bought, but earlier ones will have to be repaired or remade individually.

Leathercloth was also used for the rear parcels shelf of Mk II Coupés and of all subsequent cars. This tends to discolour and split in the sunlight. It can of course be replaced, and shaped panels are available from specialists – although the panels currently on the market come without the network of punched holes for the radio speaker. Anyone who wishes to punch his or her own holes in exactly the right places might like to take a couple of days' holiday and to have a helper on hand ready to provide plenty of tea and sympathy! Carpet and wood trim may have some fairly obvious maladies, and the pitfalls of replacing or refurbishing these have already been discussed.

The instrument panel and controls do not usually present difficulties, although it is quite common to find splits in the main spokes of the steering wheel, usually concealed under the horn ring. The horn ring itself often rattles, and this is an irritating problem which can be difficult to cure. There may also be splits in the steering column shroud. Replacements for these items can only be found second-hand, and examples of the grey steering wheels and column shrouds used on the earliest cars are very hard to come by. Broken switches on the early cars will also be hard to replace.

Mechanical Elements

Strangely enough, mechanical problems are in most cases the least expensive and the easiest to solve on both the six-cylinder and the eight-cylinder cars. Axle troubles are very rare, and the worst that is likely to happen to the springs is that the front ones will need to be reset (by turning them in their mountings) while the rears will need to be retempered. The Metalastik bushes on the back mountings of the rear springs may also deteriorate. When new, all springs were protected by gaiters made of black cloth which were held on by plastic ties, but few cars still have the original items in place.

Bushes wear in the front suspension, however, and it is worth checking the state of those on the anti-roll bar (both the ones on the subframe and the ones at the top of the swivel pins). Front suspension ball-joints on the later cars were sealed for life, but the original joints have in most cases exceeded their life expectancy by now. Replacement ball-joints are expensive and are not always readily available. The power steering box, where fitted, may also leak – and these leaks may be difficult to trace. Other fluid leaks from the power-assisted steering system are generally easier to repair, as they are caused by worn hoses or poor sealing at joints.

Brakes have never been a particular problem on P5s of any vintage, although the all-drum system of the first 3-litres is marginal even when performing at its best. Disc brakes may seize on a car which has been in storage for a long time, and of course the handbrake linkage can also seize. The problem usually lies in the rod linkage over the rear axle, although it is not unknown for the cable to fray and break – most commonly in conjunction with some other problem affecting the rear brakes. Brake servos can of course give trouble, and will lose their efficiency and eventually cease to function when internal rubber seals become porous. Getting at the servo on 3-litre models is difficult, as it is mounted on the bulkhead underneath the steering column; the job is far easier on V8-engined cars.

Gearboxes do not generally give much trouble, although when checking an overdrive 3-litre it is well worth ensuring that the overdrive does engage and disengage as it should do: seizures are fairly common on cars which do not see regular use. The layshaft bearings can give way on high-mileage manual gearboxes, and make a rattling noise at idle when the clutch is depressed. First gear has a characteristic vintage whine which is not indicative that there is a problem.

As for the automatics, the later Borg Warner type 35 in the Mk III 3-litres and P5Bs has a better change quality than the older DG type, but even so is not as smooth as more modern designs of transmission. Yet upchanges and downchanges should always be clean, and any hesitation or reluctance to change points to some sort of a problem, even if it is nothing worse than a low fluid level in the gearbox. In the V8-engined cars, first gear always whines a little, and slow-speed changes from second down to first often produce an unnerving 'clonk' from the transmission; these were probably the main reasons why Rover insisted on a first gear lock-out facility.

Lastly, there are the engines. The six-cylinders are robust and extremely refined pieces of machinery, and any serious roughness will warn of a problem. Some owners adjust the carburettor to run as lean as possible, hoping that they will save a little on fuel in the process. Sadly, the usual result is that number six exhaust valve (the one nearest the bulkhead) burns. Tappets need to be adjusted on a regular basis, and as this is not a quick job (it also costs money, as the rocker cover gasket has to be replaced) it is often neglected. The characteristic rattling of maladjusted tappets on these engines is familiar also to owners of the later P4 models and certain Land Rovers, with which the 3-litre shares its basic engine design. These engines were designed to use oil, and a certain amount of blue smoke in the exhaust is normal. Oil consumption – something the owners of modern cars have completely forgotten about – can go as high as one pint every 250 miles or so before bore wear need be counted as serious. The gentle winking of the oil pressure warning light at idle is best considered as one of the car's charming features and need not give rise to worries about wear unless there are other indications which would also suggest that.

The engine bay of the P5 is wide enough to allow plenty of room for working around either the six-cylinder or the eight-cylinder engine. This is the engine bay of a P5B, showing the distinguishing tubular members of the subframe adapted to take the V8 engine.

Spares for the 3-litre engines are not generally a problem, and owners of the 2.6-litre type will find that most parts for their engines are shared with the contemporary P4 six-cylinder engine. Those who have one of the rare 2.4-litre engines may well encounter spares problems, however. It is worth pointing out that a full engine overhaul on any one of the six-cylinder types is likely to prove very expensive.

The V8 engine shares its reputation for robustness with its six-cylinder predecessor. Its oil pressure at idle is very low, however, and may cause the oil warning light to come on when the car is stationary – particularly if the engine is straining against the torque converter. There is nothing to worry about if the light goes off as soon as the accelerator is depressed. Nevertheless, V8 engines do thrive on regular oil changes, and it is advisable to change the oil every 3,000 miles on a car which is not used much. Dirty oil tends to block up the oilways in the cylinder head, and oil starvation at the top end leads to noisy rockers and valve-gear. The hydraulic tappets also depend on a good supply of clean oil to function properly, and they will also become noisy if they are unable to pump up properly or unable to rotate within their housings.

Tuning problems may also afflict the V8s. The engines were designed to run on

187

100-octane leaded petrol, which became unavailable many years ago. Retarding the ignition timing allows the cars to be run on 98-octane fuel, but when the Rover engineers recommended this for touring in countries where 100-octane fuel was not available, they also recommended limiting the maximum speed to around 90mph. Not all V8s are happy with the same amount of ignition retard; the spread of manufacturing tolerances and the degree of wear both have an influence on the outcome, and some V8s simply cannot be persuaded to idle smoothly with a retarded ignition. Rough idling can of course also be caused by out-of-balance carburettors, although this is most likely to be the problem only when other maladies – such as acceleration flat spots – are present. A V8 in good health should be extremely smooth all the way through its rev range.

Unleaded fuel was simply not an issue when the P5Bs were new, and no-one has yet produced a definitive answer to the question of whether unleaded fuel will eventually cause valve seat damage in a P5B's engine. Rover *did* announce some years ago that all V8 engines built after 1970 had hardened valve seats suitable for use with unleaded fuel, and there does not appear to have been a change of valve seat material associated with that date. In theory, then, all the V8 engines built after 1967 should be safe with unleaded fuel, but no-one has yet proved that. Worth thinking about is that an engine designed to run on 100-octane leaded fuel may not take kindly to the twin changes of a retarded ignition to suit 98-octane fuel and the use of unleaded petrol.

Three other V8 problems are worth mentioning. The first is that the rear main bearing oil seal may develop a leak, and is next to impossible to replace with the engine in the car. The job *can* be done with the aid of a special tool known as a Chinese finger, but it is not an enjoyable experience! Some owners have taken the more expensive route of having the cylinder block machined to accept the later lipped type of oil seal, which does effect a permanent cure. The second problem is that the engines sometimes develop a ticking sound from the top end which is extremely hard to trace. If it gets louder on acceleration as well as more frequent, the chances are that it is caused by a leak between one of the exhaust manifolds and the cylinder block. The fixing bolts are held in place by lockwashers, but these are sometimes less effective than they might be, leaving the bolt free to come loose and cause a gas leak at the joint.

The third problem centres on the notorious AED or automatic choke. Rover never did manage to make this work faultlessly, and it tends to stick in the 'on' position or not to work at all. Starting troubles and stalling are typical of the consequences. Most cars have now been converted to manual choke, but it is possible to convert the AED to electrical operation and switch it on and off from the dashboard. A few cars have also been modified in this way.

THE CLUBS AND THE SPECIALISTS

Ownership of a P5 or P5B is immeasurably enhanced by membership of one of the clubs which cater for the model. Outside Britain, where the cars are less plentiful, the Rover clubs tend to embrace all models, and sometimes Land Rovers as well. There is not room to include all their addresses here, but they do advertise in the enthusiast car magazines of their native countries.

In Britain, there are two main clubs and a number of regional associations. The club which caters for Rovers of all ages is the

Rover Sports Register,
c/o Cliff Evans,

8 Hilary Close,
Great Boughton,
Chester CH3 5QP.

The specialist club, which is devoted to the P5 and P5B models alone, and also arranges for the remanufacture of certain hard-to-find items, is the

Rover P5 Owners' Club,
c/o Geoff Moorshead,
13 Glen Avenue,

Ashford,
Middlesex TW15 2JE.

Both clubs produce bimonthly magazines which help members to keep in touch, give details of meetings and rallies, offer technical advice and also list cars and spares for sale. These magazines are also the best way of finding out who the specialist spares suppliers and restorers are and how to get in touch with them.

Performance figures

The truth is that most modern cars with engines one-third the size of those in the P5s and P5Bs will accelerate faster, corner better and return very much better fuel economy. That said, it is interesting to compare the performance and fuel-consumption figures for these Rovers. In all cases, the figures are those supplied by the Rover Company when the cars were new. Road-test figures printed in the motoring magazines did not always agree with the figures which Rover obtained, although the official figures were reasonably realistic. These figures might influence your choice of P5!

Mk I and Mk IA models

Performance	*Overdrive*	*Automatic*
0–60mph	16 seconds	17.5 seconds
30–50mph in top gear	9 seconds	10.5 seconds
50–70mph in top gear	12 seconds	14 seconds
Top speed	97.5mph	97.5mph
Fuel consumption	18–24mpg	18–24mpg

Mk II and Mk III models

Performance	*Overdrive*	*Automatic*
0–60mph	15 seconds	16.1 seconds
30–50mph in top gear	8.9 seconds	not applicable
50–70mph in top gear	10.5 seconds	12.7 seconds
Top speed	111.7mph (Saloon)	
	115.3mph (Coupé)	106mph
Fuel consumption	18–24mpg	18–24mpg

3.5-litre and 3½-litre models

0–60mph	12.14 seconds (in D1 or D)
	14.46 seconds (in D2 or 2 and D)
30–50mph in top gear	Not quoted
50–70mph in top gear	Not quoted
Top speed	115mph
Fuel consumption	18.2mpg (quoted as 17.6mpg in 1968-season literature)

Index